SpringerBriefs in Applied Sciences and Technology

More information about this series at http://www.springer.com/series/8884

Lioz Etgar

Hole Conductor Free Perovskite-based Solar Cells

 Springer

Lioz Etgar
Institute of Chemistry, Casali Center for
 Applied Chemistry, the Harvey M. Kruger
 Family Center for Nanoscience and
 Nanotechnology
The Hebrew University of Jerusalem
Jerusalem
Israel

ISSN 2191-530X   ISSN 2191-5318   (electronic)
SpringerBriefs in Applied Sciences and Technology
ISBN 978-3-319-32989-5   ISBN 978-3-319-32991-8   (eBook)
DOI 10.1007/978-3-319-32991-8

Library of Congress Control Number: 2016939937

Printed on acid-free paper

This Springer imprint is published by Springer Nature
The registered company is Springer International Publishing AG Switzerland

Preface

This brief book results from the extensive work that has been done in my laboratory in the last 3 years.

As will be discussed in this brief book, a breakthrough occurs in the photovoltaic (PV) field in the last 4 years. A new material called organo-metal halide perovskite (OMHP) entered the PV field. To be completely correct, the OMHP is not a new material and already in 1990s researchers around the world (mainly from Japan) worked on characterizing this material. But the main progress related to the PV field was made at 2012 as described in more detail in this book. This year was my last year as a postdoctoral researcher in Prof. Gratzel laboratory where I was one of the pioneering researchers working with this exciting material, which results in an early publication on the use of the OMHP as light harvester and hole conductor at the same time in the solar cell.

Starting 2012 as established my current research group which have developed further the OMHP as a material and in a PV cell.

Even though just 3–4 years past from the main breakthrough with this material, I think it is the right time to summarize the basic fundamental properties and some of its exciting abilities as a fascinating material for optoelectronic applications.

This book mainly discusses our discovery that the OMHP can function as hole conductor (HTM) and light harvester in the solar cell at the same time as so-called HTM-free perovskite solar cells. It brings the ability to tune the OMHP properties (e.g., optical, physical, and electronic) and the use of the OMHP in different solar cells structures as we demonstrating the advantageous of this material on the solar cell properties.

I would like to express my appreciation to all of my students at the Hebrew University of Jerusalem who worked intensively with passion on this topic; without them, we could not make such an influenced contribution to the field. I also thank Mayra Castro from springer publisher for excellent collaboration. Finally, I would like to thank my wife and my three children for their support, happiness, and love during these years.

Contents

Chapter 1
Organo-Metal Lead Halide Perovskite Properties

1.1 Perovskite Crystal Structure

The inorganic perovskite compounds were discovered in 1839 and named after the Russian mineralogist L.A. Perovski, who first characterized the structure of perovskite compounds with general crystalline formula of ABX_3. Figure 1.1 shows the basic structural arrangement, where X is anion (oxygen or halogen), A is a bulky cation which occupies a cubo-octahedral site shared with 12X anions, and B is a smaller cation stabilized in an octahedral site shared with 6X anions. Large numbers of inorganic perovskite oxides have been extensively studied due to their electrical properties of ferroelectricity or superconductivity. Halide perovskite received their attention when Mitzi et al. [1] discovered that layered organo-metal halide perovskites exhibit semiconductor-to-metal transition with increasing dimensionality. Moreover, the band gap energy decreased with increasing dimensionality, which is suitable for photovoltaic applications.

The organo-lead halide perovskite is obtained when in a classical perovskite compound, the A site is replaced by an organic cation, which is often methyl ammonium ($CH_3NH_3^+$) or formamidinium ($NH_2–CH_2=NH_2^+$). B is usually Pb^{2+}, and X is either I^-, Cl^-, or Br^- anions. The large organic cation groups balance the charge between the octahedron layers in the 3D network (Fig. 1.1).

The formability of the 3D perovskite structure can be estimated by Goldschmidt tolerance factor (t) and an octahedral factor (μ) [2]. The tolerance factor is the ratio of the (A–X) distance to the (B–X) distance. And is given by $= \frac{(R_A + R_B)}{\sqrt{2}(R_B + R_X)}$. The octahedral factor is defined as $\mu = R_B/R_X$ where R_A, R_B, and R_X are the ionic radii of the corresponding ions at the A, B, X sites, respectively. Formability study of alkali metal halide perovskite determined that ideal cubic structure was stabilized when $0.813 < t < 1.107$ and $0.442 < \mu < 0.895$ [2]. A smaller t value could lead to a lower symmetry tetragonal or orthorhombic structure, whereas larger t value can destabilize the 3D network, creating a 2D structure [3]. Taking $MAPbI_3$, the radii of $MA^+ = 180$ pm, $Pb^{2+} = 119$ pm, $I^- = 220$ pm, thus the t factor was calculated to

© The Author(s) 2016
L. Etgar, *Hole Conductor Free Perovskite-based Solar Cells*,
SpringerBriefs in Applied Sciences and Technology,
DOI 10.1007/978-3-319-32991-8_1

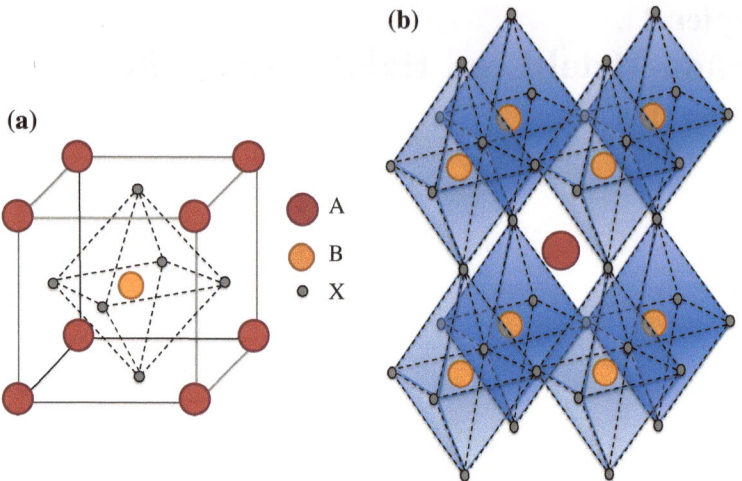

Fig. 1.1 **a** Ball and stick model of the basic cubic perovskite structure and **b** their extended network structure connected by the corner-shared octahedral

be 0.83 and μ factor is 0.54 [4] as a result, MAPbI$_3$ is expected to have a cubic structure. However, phase transition in MAPbX$_3$ also occurs when temperature is varying. At temperature below 162 K MAPbI$_3$ forms in an orthorhombic phase, by increasing the temperature up to 327 K it transforms to tetragonal phase, and as temperature increase above 327 K, MAPbI$_3$ undergoes transition phase to the cubic phase [4]. Therefore, at room temperature MAPbI$_3$ holds tetragonal structure.

1.2   Perovskite Energy Level

Organo -metal halide perovskite compounds have wide-direct band gaps, which can be tuned by either changing the alkyl group, the metal atom or the halide [5, 6].

Kim et al. studied the direct band gap of tetragonal MAPbI$_3$, the optical band gap (E_g) was determined using diffuse reflectance spectroscopy, and the valance band maximum (E_{VB}) was determined using Ultraviolet photoelectron spectroscopy (UPS).

The optical band gap of MAPbI$_3$ deposited on mesoporous TiO$_2$ was found to be 1.5 eV, and the valance band energy position was estimated to be −5.43 eV below vacuum level using UPS [7], which is consistent with the previous report [8]. From the optical band gap and the position of the valence band, the conduction band (E_{CB}) was estimated to be −3.93 eV, as the energy band gap of MAPbBr$_3$ reported to be 2.2 eV [9], and for MAPbCl$_3$ 3.11 eV [10]. Moreover, phase transformation due to change in temperature also affect the bang gap. Lower symmetry of orthorhombic increases the optical gap to 1.6 eV, and higher symmetry of the cubic phase decreases the optical band gap to 1.3 eV [4]. Reducing the optical band gap is possible by replacing the organic group with a larger group, such as $NH_2-CH_2=NH^+_2$, or by replacing the lead with tin (Sn) [11–13].

1.3 Absorption Coefficient of Perovskite

Organo-lead halide perovskites have drawn substantial interest as a light harvester due to their absorption onset of more than 800 nm in the visible spectrum, and due to their large absorption coefficient [3, 14–16] that is crucial for PV application. Perovskite showed an absorption coefficient that is 10 times greater than that of the N719 conventional dye molecule used so far in DSSC. The absorption coefficient of $MAPbI_3$ at 550 nm wavelength is 1.5×10^4 cm^{-1}, indicating penetration depth of 0.66 μm [14]. The absorption coefficients of perovskite versus silicon and GaAs was compared, [3] presenting that perovskite has much higher absorption coefficient than silicon and GaAs due to its direct band gap and its higher density of state.

1.4 Balance Charge Transport of Perovskite

In addition for having high absorption coefficient, Perovskite is also characterized by efficient electron and hole transport properties. Due to its balance charge transport property, fabrication of solar cell without hole transport material and mesoporous TiO_2 with high efficiencies is possible, which is impossible in conventional dye sensitized solar cells.

The electron diffusion length for $MAPbI_3$ was estimated as 130 nm, and the hole diffusion length was estimated as 100 nm, For $MAPbI_{3-x}Cl_x$ the electron diffusion length of electron was about 1069 nm, while the hole diffusion length was about 1213 nm [16, 17]. The recombination time of the electron and hole is very slow, in tens of microseconds. Moreover, charge accumulation properties were identified for perovskite. The result indicated the existence of high-density state and supported the weakly bounded excitons in perovskite, which can lead to a high open-circuit voltage in perovskite solar cells [18].

References

1. Mitzi DB, Feild CA, Harrison WTA, Guloy AM (1994) Conducting tin halides with a layered organic-based perovskite structure. Nature 369:467–469
2. Li C, Lu X, Ding W, Feng L, Gao Y, Guo Z (2008) Formability of ABX3 (X = F, Cl, Br, I) halide perovskites. Acta Crystallogr B 64:702–707
3. Yin WJ, Yang JH, Kang J, Yan Y, Wei S-H (2015) Halide perovskite materials for solar cells: a theoretical review. J Mater Chem A 3:8926–8942

4. Liu X, Zhao W, Cui H, Xie Y, Wang Y, Xu T, Huang F (2015) Organic–inorganic halide perovskite based solar cells—revolutionary progress in photovoltaics. Inorg Chem Front 2:315–335

5. Mitzi DB (2000) Templating and structural engineering in organic–inorganic perovskites. J Chem Soc, Dalton Trans 1:1–12

6. Knutson JL, Martin JD, Mitzi DB (2005) Tuning the band gap in hybrid tin iodide perovskite semiconductors using structural templating. Inorg Chem 44:4699–4705

7. Kim H-S, Lee C-R, Im J-H, Lee K-B, Moehl T, Marchioro A, Moon S-J, Baker R-H, Yum J-H, Moser JE, Grätzel M, Park N-G (2012) Lead iodide perovskite sensitized all-solid-state submicron thin film mesoscopic solar cell with efficiency exceeding 9 %. Sci Rep 2:591

8. Schulz P, Edri E, Kirmayer S, Hodes G, Cahen D, Kahn A (2014) Interface energetics in organo-metal halide perovskite-based photovoltaic cells. Energy Environ Sci 7:1377–1381

9. Noh JH, Im SH, Heo JH, Mandal TN, Seok SI (2013) Chemical management for colorful, efficient, and stable inorganic–organic hybrid nanostructured solar cells. Nano Lett 13:1764–1769

10. Kitazawa N, Watanabe Y, Nakamura Y (2002) Optical properties of CH3NH3PbX3 (X = halogen) and their mixed-halide crystals. J Mater Sci 37:3585–3587

11. Pang S, Hu H, Zhang J, Lv S, Yu Y, Wei F, Qin T, Xu H, Liu Z, Cui G (2014) NH2CH⁄ 4NH2PbI3: an alternative organolead iodide perovskite sensitizer for mesoscopic solar cells. Chem Mater 26:1485–1491

12. Im J-H, Chung J, Kim S-J, Park N-G (2012) Synthesis, structure, and photovoltaic property of a nanocrystalline 2H perovskite-type novel sensitizer (CH3CH2NH3)PbI3. Nanoscale Res Lett 7:353

13. Hao F, Stoumpos CC, Chang RPH, Kanatzidis MG (2014) Anomalous band gap behavior in mixed Sn and Pb perovskites enables broadening of absorption spectrum in solar cells. J Am Chem Soc 136:8094–8099

14. Im J-H, Lee CR, Lee J-W, Park S-W, Park N-G (2011) 6.5 % efficient perovskite quantum-dot-sensitized solar cell. Nanoscale 3:4088–4093

15. Kojima A, Ikegami M, Teshima K, Miyasaka T (2012) Highly luminescent lead bromide perovskite nanoparticles synthesized with porous alumina media. Chem Lett 41:397–399

16. Xing G, Mathews N, Sun S, Lim SS, Lam YM, Grätzel M, Mhaisalkar S, Sum TC (2013) Long-range balanced electron- and hole-transport lengths in organic–inorganic CH3NH3PbI3. Science 342:344–347

17. Stranks SD, Eperon GE, Grancini G, Menelaou C, Alcocer MJ, Leijtens T, Herz LM, Petrozza A, Snaith HJ (2013) Electron-hole diffusion lengths exceeding 1 micrometer in an organometal trihalide perovskite absorber. Science 342:341–344

18. Kim H-S, Mora-Sero I, Pedro VG, Santiago FF, Juarez-Perez EJ, Park N-G, Bisquert J (2013) Mechanism of carrier accumulation in perovskite thin-absorber solar cells. Nat Commun 4:2242

Chapter 2
The Evolution of Perovskite Solar Cells Structures

2.1 Perovskite Solar Cells with Liquid Electrolyte

Organo-metal lead halide perovskite was first used in solar cells in 2009 by Miyasaka et al. [1] $CH_3NH_3PbX_3$ (X = Br, I) was applied as a sensitizer in a Dye Sensitized Solar Cell (DSSC) with liquid electrolyte. The conductive glass was coated with TiO_2 NPs and the perovskite was deposited on top the TiO_2 NPs, the cell structure is depicted in Fig. 2.1. PCE of 3.1 % was demonstrated for X = Br and 3.8 % for X = I [1], with spectral response covering the whole visible region till 800 nm. Park et al. [2] reported in 2011 PCE of 6.5 % employing similar structure in which 2–3 nm sensitized nanoparticles of $CH_3NH_3PbI_3$ perovskite was deposited in a one-step solution process on top 3.6 μm thick TiO_2 NPs with iodide/iodine-based redox electrolyte. They have showed that perovskite exhibited better absorption than the standard N719 dye sensitizer, but the ionic crystal perovskite dissolved in the polar electrolyte, resulting in rapid degradation of performance [2].

The effect of TiO_2 film thickness was investigated in sensitized $CH_3NH_3PbI_3$ perovskite solar cells [3] studying the charge transport, recombination, and PV performance using iodide-based electrolyte. It was found that $CH_3NH_3PbI_3$ is relatively stable in a nonpolar solvent with low iodide concentration. However, polar electrolyte or higher iodide concentration of nonpolar electrolyte achieved higher cell performances, but result in degradation or bleaching of the perovskite.

2.2 Perovskite Solar Cells with Solid-State Hole Conductor

Replacing the liquid electrolyte with a solid-hole conductor material (HTM) solved the immediate instability problem of the perovskite-based solar cells and increased their efficiencies. Park and Gratzel [4] reported in 2012 of a solid-state stable

© The Author(s) 2016
L. Etgar, *Hole Conductor Free Perovskite-based Solar Cells*,
SpringerBriefs in Applied Sciences and Technology,
DOI 10.1007/978-3-319-32991-8_2

Fig. 2.1 Structure of liquid-sensitized perovskite solar cell. The *brown* color indicates the perovskite

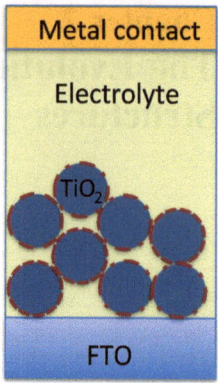

perovskite solar cell. They deposited $CH_3NH_3PbI_3$ onto a mesoporous TiO_2 film and introduced a spiro-MeOTAD (2,2′,7,7′-tetrakis(N,N-di-p-methoxyphenylamine)-9,9′-spirobifluorene), HTM. When dissolved in an organic solvent, spiro-MeOTAD percolates the TiO_2 NPs, leaving only solute molecules after solvent evaporation (Fig 2.2). The replacement of the liquid electrolyte with spiro-MeOTAD not only improved the stability of the solar cell, but also boosted its efficiency to 9.7 % [4].

Snaith et al. [5] reported a PCE of 10.9 % with high V_{oc} of 0.98 V, in which mixed halide perovskite $CH_3NH_3PbI_{3-x}Cl_x$ coated mesoporous Al_2O_3 film rather TiO_2 film with spiro-MeOTAD as HTM. It was shown that the mesoporous Al_2O_3 served only as a scaffold for the pervoskite and not as an electron acceptor as in the case of mesoporous TiO_2, due to the higher valance band of Al_2O_3. As a result, the electrons must remain in the perovskite, and transferred through the perovskite to the compact TiO_2 then to the FTO electrode. They were able to demonstrate that perovskite is able to transport both electrons and holes between cell terminals, besides of being used as a sensitizer [5]. Their cell structured named "meso-superstructured solar cell" (MSSC), because of the existence meso-structure and the absence of electron injection to meso-structure.

Fig. 2.2 Structure of solid-sensitized perovskite solar cell. BL-blocking layer; HTM-hole transport material

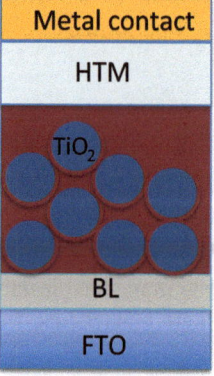

Few months later further progress was reported. The two-step deposition process was reported when the perovskite layer was prepared by spin-coating PbI_2 solution onto the electrode coated with the TiO_2 scaffold and then dip the electrode into the cation iodide solution (such as CH_3NH_3I solution). Better coverage and few hundred-nanometer perovskite over layer were achieved by this deposition technique. As a result PCE of 15 % was reported using this deposition technique.

2.3   Perovskite Solar Cells in Planar Configuration

Due to the superior properties of the perovskite, it was shown that the perovskite could function as light harvester in planar configuration. The initial planar perovskite solar cell was fabricated by Snaith et al. [5], where $CH_3NH_3PbI_{3-x}Cl_x$ was deposited through a solution process on top of FTO conductive glass coated with compact TiO_2 (BL), the cell structure is depicted in Fig. 2.3. In the beginning Low PCE of 1.8 % was achieved, due to the need of homogenous high surface coverage. Snaith and coworkers [6] then optimized the film formation of $CH_3NH_3PbI_{3-x}Cl_x$ perovskite by controlling the atmosphere, and annealing temperature and achieved PCE of 11.4 % with J_{sc} of 20.3 mA/cm^2 and V_{oc} of 0.89 V. The use of vapor-deposition technique [7] for the perovskite has improved the PCE of the planar configuration to 15.4 % with V_{oc} of 1.07 V. These findings showed that perovskite could be used in a simple architecture solar cell without the need of mesoporous n-type semiconductor. Kelly and colleague [8] improved the PCE to 15.7 % by deposition $MAPbI_3$ in a two-step deposition on top of compact ZnO instead of compact TiO_2. Yang's groups [9] demonstrated further increase in PCE to 19.3 % by modifying the commonly planar configuration.

Fig. 2.3 Structure of planar solid-state perovskite solar cell

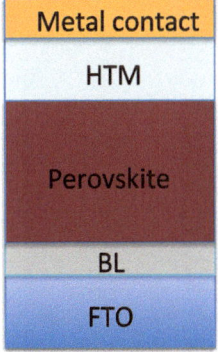

References

1. Kojima A, Teshima K, Shirai Y, Miyasaka T (2009) Organometal halide perovskites as visible-light sensitizers for photovoltaic cells. J Am Chem Soc 131:6050–6051
2. Im J-H, Lee CR, Lee J-W, Park S-W, Park N-G (2011) 6.5 % efficient perovskite quantum-dot-sensitized solar cell. Nanoscale 3:4088–4093
3. Zhao Y, Zhu K (2013) Charge transport and recombination in perovskite (CH3NH3)PbI3 sensitized TiO2 solar cells. J Phys Chem Lett 4:2880–2884
4. Kim H-S, Lee C-R, Im J-H, Lee K-B, Moehl T, Marchioro A, Moon S-J, Baker R-H, Yum J-H, Moser JE, Grätzel M, Park N-G (2012) Lead iodide perovskite sensitized all-solid-state submicron thin film mesoscopic solar cell with efficiency exceeding 9 %. Sci Rep 2:591
5. Lee MM, Teuscher J, Miyasaka T, Murakami TN, Snaith HJ (2012) Efficient hybrid solar cells based on meso-superstructured organometal halide perovskites. Science 338:643–647
6. Eperon GE, Burlakov VM, Docampo P, Goriely A, Snaith HJ (2014) Morphological control for high performance, solution-processed planar heterojunction perovskite solar cells. Adv Funct Mater 24:151–157
7. Liu M, Johnston MB, Snaith HJ (2013) Efficient planar heterojunction perovskite solar cells by vapour deposition. Nature 501:395–399
8. Liu D, Kelly TL (2013) Perovskite solar cells with a planar heterojunction structure prepared using room-temperature solution processing techniques. Nat Photon 8:133–138
9. Zhou H, Chen Q, Li G, Luo S, Song TB, Duan H-S, Hong Z, You J, Liu Y, Yang Y (2014) Interface engineering of highly efficient perovskite solar cells. Science 345:542–546

Chapter 3
Hole Transport Material (HTM) Free Perovskite Solar Cell

Abstract At the same time of the perovskite discovery to function as efficient light harvester in the solar cell, Etgar et al. [1] first proposed a heterojunction device structure of FTO/TiO$_2$/CH$_3$NH$_3$PbI$_3$/Au in which CH$_3$NH$_3$PbI$_3$ was used as a p-type semiconductor and 500 nm mesoscopic TiO$_2$ was used as an n-type semiconductor.

3.1 Introduction

At the same time of the perovskite discovery to function as efficient light harvester in the solar cell, Etgar et al. [1] first proposed a heterojunction device structure of FTO/TiO$_2$/CH$_3$NH$_3$PbI$_3$/Au in which CH$_3$NH$_3$PbI$_3$ was used as a p-type semiconductor and 500 nm mesoscopic TiO$_2$ used as an n-type semiconductor. They demonstrated that CH$_3$NH$_3$PbI$_3$ could act as light harvester as well as a hole transporter, achieving initial PCE of 5.5 %. This structure simplifies the device structure, enhances the stability of solar cell, eliminates the use of (Hole Transport Material) HTM that reduces the solar cell cost, and eliminates HTM infiltration issues in mesoporous solar cells.

Following this pioneering work it was found that perovskite has the ability to conduct both electrons and holes, which is supported by its long diffusion length of both carriers [2, 3], it was found reasonable to make a p–n junction HTM-free perovskite solar cell.

Etgar et al. [4] were able to improve the PCE for HTM-free perovskite by using a 300 nm mesoporous TiO$_2$ film, and tuning the deposition technique of the perovskite layer. Capacitance–Voltage measurements and Mott-Schottky analysis confirmed the existence of a depletion region between the mesoporous TiO$_2$ and the CH$_3$NH$_3$PbI$_3$. The depleted HTM-free TiO$_2$/CH$_3$NH$_3$PbI$_3$ heterojunction solar cell demonstrated PCE of 8.1 %. Figure 3.1 shows the scheme of the depleted TiO$_2$/CH$_3$NH$_3$PbI$_3$ heterojunction solar cell and its energy level diagram. The PCE further improved to 10.85 % with finding the correlation of the depletion region width at the TiO$_2$/CH$_3$NH$_3$PbI$_3$ Junction and PCE, while using the two-step deposition technique [5].

© The Author(s) 2016
L. Etgar, *Hole Conductor Free Perovskite-based Solar Cells*,
SpringerBriefs in Applied Sciences and Technology,
DOI 10.1007/978-3-319-32991-8_3

Fig. 3.1 a Schematic
illustration of the depleted
heterojunction HTM-free
perovskite solar cell.
b Energy level diagram
showing the depletion layer
between TiO_2 and
$CH_3NH_3PbI_3$. Taken with
permission from Ref. [4]

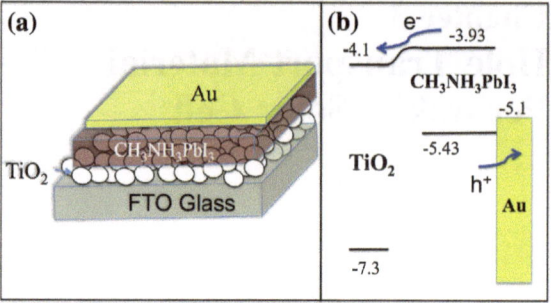

3.2   Mesoporous HTM-Free Perovskite Solar Cell

As explained above the perovskite can be functioned as light harvester and hole
conductor in the solar cell where it is deposited on top of a mesoporous film.

A two-step deposition technique was used to optimize the perovskite deposition
and to enhance the solar cell efficiency.

The first step was spinning PbI_2 on the mesoporous TiO_2 film and annealing at
70 °C, while the second step was dipping the PbI_2 electrode into CH_3NH_3I solution. After the PbI_2 reacts with the CH_3NH_3I, the $CH_3NH_3PbI_3$ was formed. An
important step in this deposition technique is the wait time period after dropping the
PbI_2 on the mesoporous TiO_2 film before spinning. We investigated the wait time
parameter and observed its influence on the PV parameters of these cells. The
results are summarized in Fig. 3.2. The highest efficiency (8.0 %) was achieved at a
wait time of 3 min. During 3 min of waiting time, the PbI_2 creates a uniform
coating on the mesoporous TiO_2 surface while a longer waiting time (5 min) causes
evaporation of the PbI_2's solvent, which could create a nonuniform coating of the
PbI_2 on the mesoporous TiO_2 surface. Conversely, the coverage of the surface by
PbI_2 is not complete during a too short wait time.

Fig. 3.2 The current density
and the efficiency of the cells
versus the wait time
(min) between dropping the
PbI_2 until the beginning of the
spinning. Reproduced from
Ref. [5] with permission from
the PCCP Owner Societies

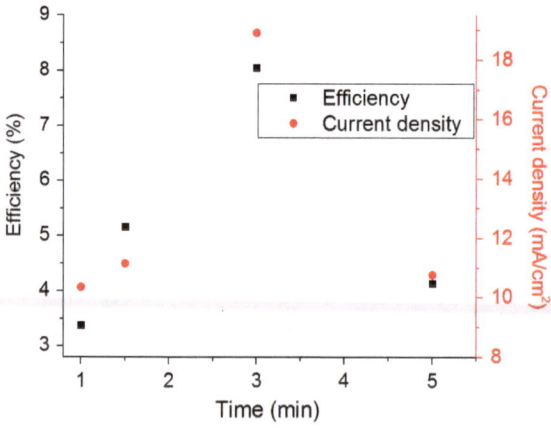

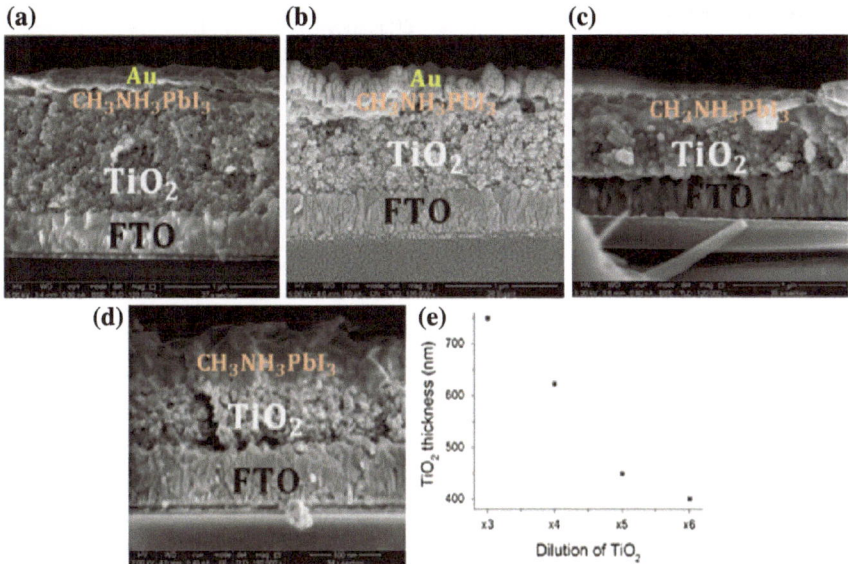

Fig. 3.3 XHR-SEM of the different TiO_2 thicknesses in the complete $TiO_2/CH_3NH_3PbI_3$ perovskite-based solar cells. **a** The thicker TiO_2 film corresponds to 3 times dilution, see Fig. 3.3e. **b** 4 times dilution. **c** 5 times dilution. **d** 6 times dilution. Importantly, the thickness was measured by surface profiler which its results are shown in (**e**). Reproduced from Ref. [5] with permission from the PCCP Owner Societies

The influence of the TiO_2 film thickness was investigated by making hole conductor-free perovskite-based solar cells using different thicknesses of meso-porous TiO_2 films while keeping the $CH_3NH_3PbI_3$ perovskite film thickness constant (same deposition parameters were used for the perovskite deposition). Figure 3.3a–d presents extra high-resolution scanning electron microscopy (XHR-SEM) of the various TiO_2 thicknesses in the complete set of thicknesses of $TiO_2/CH_3NH_3PbI_3$ perovskite solar cells. It can be observed that the $CH_3NH_3PbI_3$ formed an overlayer on top of the TiO_2 film. Probably some of the $CH_3NH_3PbI_3$ is penetrated into the TiO_2 film, however, the thick overlayer of the $CH_3NH_3PbI_3$ film $(300 \pm 50$ nm) indicates that most of the perovskite is staying on top of the TiO_2 film. The highest efficiency was observed for cells made with mesoporous TiO_2 film of 620 ± 25 nm thickness. In these cells, the current density and the open circuit voltage were higher than the other cells made with different thicknesses of TiO_2.

To understand the reason for the difference in PV performance observed when using a variety of TiO_2 film thicknesses, capacitance voltage measurements were performed. To estimate the depletion region width, Mott Schottky analysis on the $TiO_2/CH_3NH_3PbI_3$ heterojunction solar cells was performed.

Using Mott Schottky analysis one can calculate the W_n, W_p, total depletion width (W_t) and the built-in potential (V_{bi}) for the various TiO_2 film thicknesses. The

Fig. 3.4 Efficiency and depleted fraction of the TiO$_2$ as a function of the TiO$_2$ film thickness. Reproduced from Ref. [5] with permission from the PCCP Owner Societies

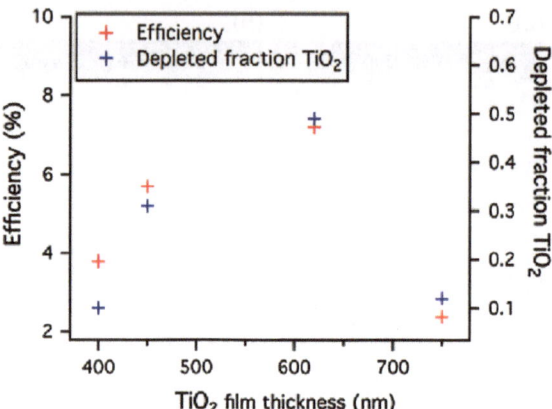

correlation between the power conversion efficiency for the various TiO$_2$ film thicknesses and the depleted fraction of the TiO$_2$ film can be observed in Fig. 3.4.

In the case of the highest efficiency, the total depletion width (W_t) is the maximum and equals 395 nm; moreover, half of the TiO$_2$ film is depleted (depleted fraction equals to 0.49), suggesting that the depletion region indeed assists in the charge separation and suppresses the back reaction, and consequently contributes to the increase in the power conversion efficiency of the cells. For the low efficiencies, the Wt is the lowest, and the depleted fraction of the TiO$_2$ is around 0.1, which means that only 10 % of the total TiO$_2$ thickness is depleted. Further, there is good agreement between the open circuit voltage and the built-in potential observed from the Mott Schottky plot. As a conclusion there is a very good agreement between the depletion region width and the PCE observed.

The highest PCE observed in this study reached 10.85 % with a fill factor of 68 %, V_{oc} of 0.84 V, and J_{sc} of 19 mA/cm^2 for HTM free solar cell.

The study above discussed the TiO$_2$/perovskite interface, however, the unique point of this solar cell structure is the direct contact between the perovskite and the metal contact (Due to the lack of hole conductor) [6].

Looking deeply on this interface anti-solvent surface treatment was carried out, reducing the surface roughness of the perovskite film.

Figure 3.5a presents a schematic illustration of the CH$_3$NH$_3$PbI$_3$ deposition process. The perovskite deposition process is based on the two-step deposition method as described earlier [7, 8] with the addition of anti-solvent (toluene) treatment on the perovskite film. We observed enhancement in the photovoltaic performance as a result of the toluene treatment discussed below. Figure 3.5b shows the X-ray diffraction (XRD) spectra of perovskite film without the toluene treatment ('Standard') and with the toluene treatment. No observable variations were recognised in the XRD spectra, which suggest that no crystallographic changes have occurred. A schematic illustration of the hole transport material free solar cell structure is presented in Fig. 3.5c.

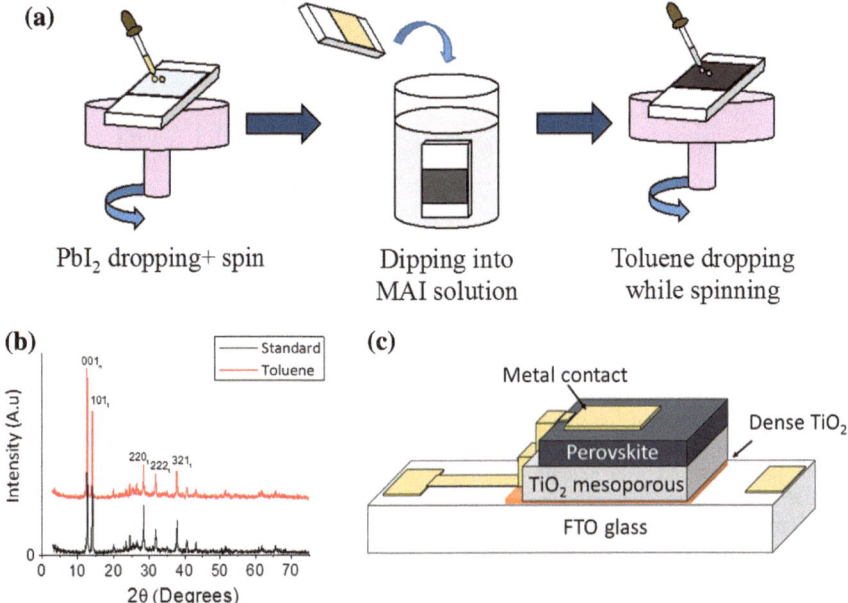

Fig. 3.5 a Schematic presentation of the anti-solvent treatment for the perovskite film deposition. **b** XRD spectra of the standard perovskite film and the perovskite film with the toluene treatment. **c** The structure of the HTM-free perovskite-based solar cell. MAI corresponds to CH_3NH_3I. Reprinted with permission from Ref. [18] Copyright (2015) American Chemical Society

The current voltage (IV) curve of the hole transport material free perovskite-based solar is presented in Fig. 3.2a with open circuit voltage (V_{oc}) of 0.91 V, fill factor (FF) of 0.65, and current density (J_{sc}) of 19 mA/cm^2 corresponding to power conversion efficiency of 11.2 %. The normalized external quantum efficiency (EQE) curve is presented at the inset of Fig. 3.6a. The integration over the EQE spectrum gives current density of 16.2 mA/cm^2 in good agreement with the J_{sc} obtained from the solar simulator. Figure 3.6b presents statistics for standard cells and for cells with toluene treatment. It is noted that the average power conversion efficiency of the standard cells is 8 ± 1 %, while the average power conversion efficiency of the toluene treated cells is 9 ± 1 %. It is clear that the toluene treatment improves the photovoltaic performance of the hole transport material-free cells. The J_{sc} and the FF change slightly while the V_{oc} is the parameter most affected by this treatment. The V_{oc} is larger by 0.05 V on average in the case of the toluene-treated cells. This enhancement can be related to the improved morphology and conductivity as discussed below. In order to prove the reliability of the results statistical analysis was preformed as described by Buriak et al. [9] First the Z score is determined according to $Z = \frac{X_1 - X_2}{\sigma/\sqrt{N}}$ where X_2 is the average PCE of the non treated cells, X_1 is the average PCE of the treated cells, σ is the standard deviation of the treated cells and N is the number of cells. Using these

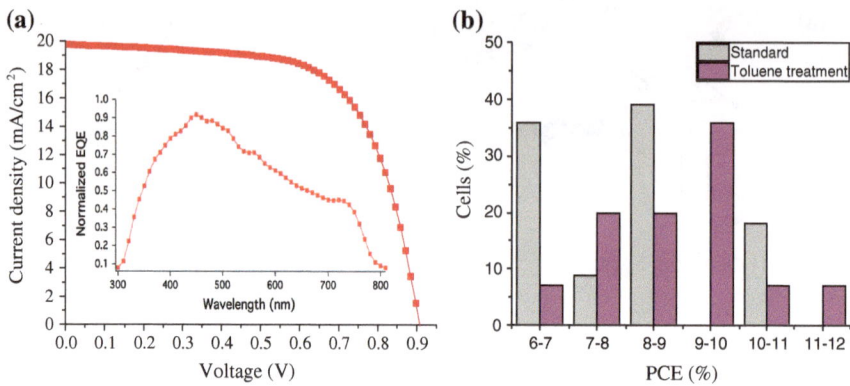

Fig. 3.6 a I–V curve of the toluene-treated hole transport material-free perovskite-based solar cell. Inset-EQE curve of the corresponding cell. **b** Statistics of the efficiency of cells without toluene treatment and with toluene treatment. The statistic analysis was performed for a total number of 28 electrodes which are equivalent to 84 cells. Reprinted with permission from Ref. [18] Copyright (2015) American Chemical Society

parameters the Z score was equal to 2.5, which correspond to p value of 0.012. Since the p value is lower than 0.05 it can be concluded with 95 % confidence that the average PCE of cells fabricated using the toluene treatment is greater than the average PCE of cells without toluene treatment. This further supports the statistical significance of the results.

To investigate the morphology and the electronical effects of the toluene treatment on the perovskite film, we performed scanning electron microscopy (SEM), atomic force microscopy (AFM), Conductive AFM (cAFM), and surface photovoltage measurements.

Top view SEM images are presented in Fig. 3.7a, d for the toluene-treated and non-treated samples, respectively. The red circles indicate the pinholes observed in the perovskite film. It can be seen that in the case of perovskite film, after the toluene treatment, there are fewer pinholes than in the perovskite film made by two-step deposition without additional treatment. It can be concluded that the perovskite coverage improved as a result of this treatment. In addition AFM was performed to observe the root mean square roughness (RMS) of the perovskite film surface. The RMS in the case of toluene-treated film is 30 nm, while for the standard (non-treated) film, the RMS is 40 nm, indicating that in addition to the better coverage achieved by the toluene treatment, the surface roughness is also smoother when treated with toluene. Since the perovskite film has a direct attachment to the metal contact in the hole transport material-free configuration, the better coverage and the lower RMS of the toluene-treated cells contribute to better photovoltaic performance. Figure 3.7b, e shows atomic force microscopy morphology and the corresponding conductive AFM (cAFM) measurements (current mapping) of the toluene-treated and non-treated $CH_3NH_3PbI_3$ perovskite films (Fig. 3.7c, f). The conductivity of the perovskite film treated with toluene is larger

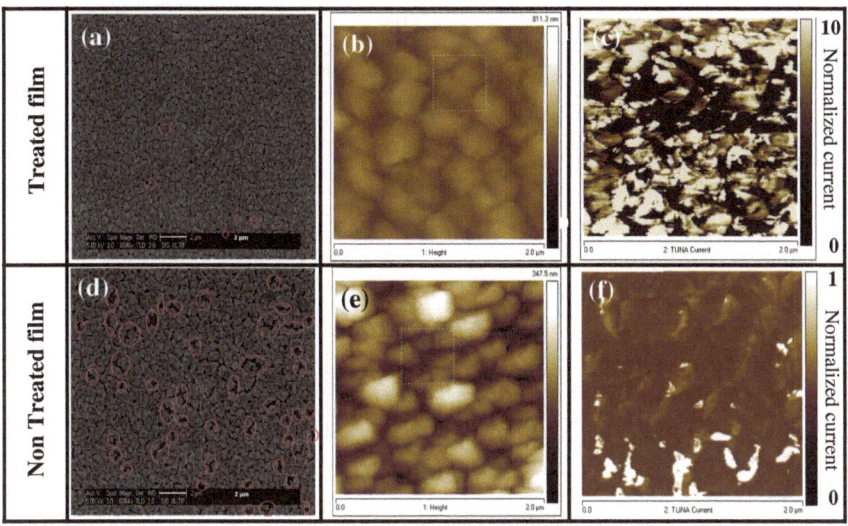

Fig. 3.7 SEM figures of **a** 'Toluene treated' solar cell and **d** 'standard' solar cell. The *red circles* indicate pinholes in the perovskite film. Morphology AFM images of **b** 'toluene treated' and **e** 'standard' solar cells. The root mean square roughness for standard cell is 40 nm and for toluene treated cell is 30 nm. Current mapping measured by c-AFM without bias. **c** Conductivity of the Toluene treated perovskite film; **f** conductivity of the non-treated perovskite film ('standard'). Reprinted with permission from Ref. [18] Copyright (2015) American Chemical Society

(by ×10 times over the conductive grains) than the non-treated perovskite film (Fig. 3.7c vs. 3.7f). No bias was applied during the conductivity measurements. In the toluene treated film, most of the grains are conductive (Fig. 3.7c), while in the non-treated perovskite film the average conductivity is much lower with slightly higher conductivity at the grain boundaries (Fig. 3.7f).

Figure 3.8a, b shows the current voltage (IV) plots of a single conductive grain, in forward and reverse scans of the non-treated and toluene-treated perovskite films, respectively. The inset of Fig. 3.8a, b presents the IV plots of a single nonconductive grain, forward and reverse scans for the two different treatments.

Several conclusions can be extracted from this measurement. (i) No hysteresis observed in the toluene treated film compared to film without toluene treatment. Clearly, the toluene treatment suppresses the hysteresis. (ii) In the case of the non-treated film (Fig. 3.8a) the cAFM measurements show direct experimental observation of the memory properties of the perovskite [10]. (iii) Looking at the IV slope in the linear region for both films (Fig. 3.8a, b), it can be observed that the slope in the IV plot is smaller in the treated perovskite than in the non-treated perovskite. The difference in the slope suggests different carrier densities of the two samples [11]. It appears that after the toluene treatment, the sample becomes more intrinsic (intrinsic semiconductor means a pure semiconductor without any significant dopant species present) than without the toluene treatment. (iv) In the case of nonconductive grains, the IV plots were almost zero (insets of Fig. 3.8a, b). From

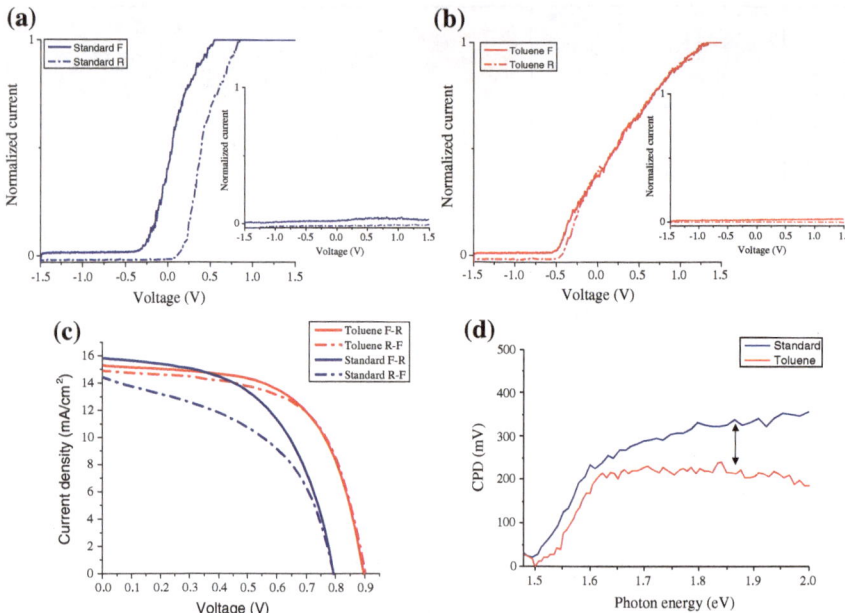

Fig. 3.8 IV plots measured on single perovskite grain using conductive AFM technique. **a** Perovskite film without toluene treatment. **b** Perovskite film with toluene treatment. **c** Current voltage curves measured by solar simulator for HTM-free perovskite solar cell. The scan rate was: 0.087 V/s. **d** Surface photovoltage measurements of toluene treated and non-treated perovskite films, the black arrow indicate the difference in the CPD. F—forward, R—reverse. Reprinted with permission from Ref. [18] Copyright (2015) American Chemical Society

the cAFM measurements, it seems that the toluene treatment does not just passivate the perovskite film, it also changes its electronic properties.

Figure 3.8c shows current voltage curves forward and reverse scans measured by solar simulator under 1 sun illumination of hole transport material-free cells, treated with toluene and non-treated. The difference between the treated cells compared to the non-treated cells is observable. The hysteresis in the non-treated cells is much more pronounced than in the toluene treated cells where a small change appears between the forward-to-reverse scan and the free reverse-to-forward scan. However, in contrast to the IV plot of the toluene-treated film measured by the cAFM, where no hysteresis was observed (Fig. 3.8b), in the current voltage measurements of the complete solar cell (Fig. 3.8c), there is still a small shift between the two scan directions. This important result suggests that the origin of the hysteresis has more than one influence, when clearly one influence on the hysteresis is related to the intrinsic properties of the perovskite, probably the memory effect [11].

To further elucidate the influence of the toluene treatment on the electronic properties of the perovskite, the surface photovoltage technique was applied on the toluene-treated and non-treated perovskite films (Fig. 3.8d). The main observation noted from the surface photovoltage spectra is the difference in the Contact

Potential Difference (ΔCPD). The Contact Potential Difference is higher by ≈100 mV (marked with an arrow in Fig. 3.8d) in the case of the standard perovskite film (non-treated film) compared to the toluene-treated film. The difference in the Contact Potential Difference suggests that the quasi-Fermi level of the toluene-treated film is higher (less negative by 0.1 eV) than the quasi-Fermi level of the non-treated toluene film. Therefore, it can be concluded that subsequently the toluene treatment, the perovskite film becomes slightly more intrinsic as also observed by the cAFM measurements. In addition, the surface photovoltage approximately indicates the band gap of the material. The band gap of the toluene-treated film extracted from the surface photovoltage spectra is 1.57 eV, while for the non-treated film, the band gap is 1.56 eV, which suggests that there is no observable change in the band gap.

Figure 3.9 illustrates the effect of the toluene treatment on the perovskite surface. We suggest that during the toluene treatment excess of halide and methylammonium ions are removed from the surface by forming a complex with the solvent similar to the previous report by Soek et al. [12]. This creates a net positive charge on the Pb atoms. Snaith et al. have reported on similar effect before the application of Lewis base passivation [13]. This interpretation is correlated to the results obtained from the cAFM. The cAFM measurement indicates that the surface after toluene treatment is more conductive than before the toluene treatment, which agrees well with the net positive charge on the Pb atoms in the case of treated perovskite surface. In addition the net positive charge of the perovskite surface after the toluene treatment is beneficial for the PV performance. Net positive charge of the perovskite surface could accept electrons more efficiently, which is useful for the interface of the perovskite with the gold contact.

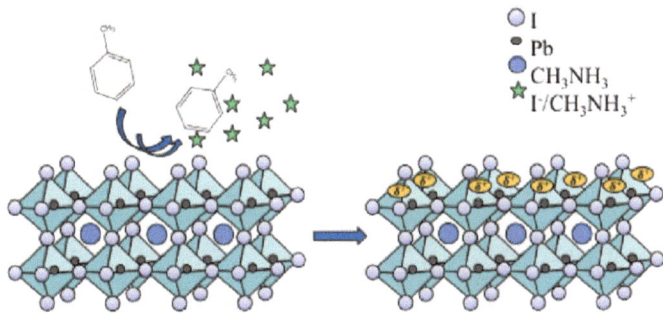

Fig. 3.9 The effect of the toluene treatment on the perovskite surface. During the toluene treatment excess of halide and methylammonium ions are removed from the surface which creates a net positive charge on the Pb atoms. Reprinted with permission from Ref. [18] Copyright (2015) American Chemical Society

3.3   Planar HTM-Free Perovskite Solar Cell [14]

As described previously the perovskite can be used in two main configurations, e.g., planar and mesoporous structures. Therefore it is interesting to have an even simpler solar cell configuration where a planar structure is used without HTM. This makes the solar cell configuration easy to fabricate and very interesting to investigate.

In this work, we present a unique planar HTM-free perovskite solar cell. The perovskite deposition was done by facile spray technique to create micron-sized grains of perovskite film. We used this deposition technique in a simple solar cell configuration consisting of FTO glass/compact TiO_2/$CH_3NH_3PbI_3$/Au. The perovskite functioned both as a light harvester and a hole conductor in this solar cell structure. The perovskite film thickness and the blocking layer thickness were varied. Interestingly, power conversion efficiency of 6.9 % was achieved for the planar HTM-free cell with 3.4 μm perovskite film thickness.

Figure 3.10a shows a schematic illustration of the spray deposition technique; the perovskite precursors are sprayed from a one-step solution onto a hot substrate. Subsequently, the solvent dimethylformamide (DMF) rapidly evaporates, immediately creating the perovskite crystals. Rapid evaporation of the solvent, which prevents percolation of the perovskite crystals into a mesoporous film, is an important stage of the deposition technique. Low photovoltaic (PV) performance is

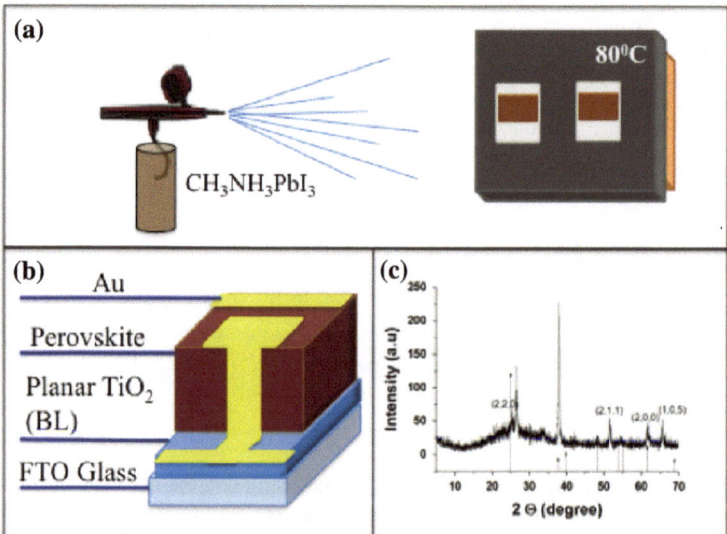

Fig. 3.10 **a** Schematic illustration of the spray deposition technique. **b** Planar HTM free perovskite solar cell structure. **c** XRD of the planar TiO_2 thin film. The anatase TiO_2 crystallographic planes are indexed in the figure. Reprinted with permission from Ref. [14] Copyright (2015) American Chemical Society

expected in the case of perovskite cells based on mesoporous metal oxide. Consequently, we used this facile deposition technique for the planar architecture as shown schematically in Fig. 3.10b. The solar cell structure composed of planar anatase TiO_2 film, as confirmed by the X-ray diffraction (XRD), shows the anatase XRD peaks in Fig. 3.10c. A thick perovskite film was deposited on the planar TiO_2 by the spray technique, following the evaporation of gold contact. The number of spray passes made over the electrode controlled the thickness of the perovskite film. In this solar cell structure, the perovskite functions as both a light harvester and a hole conductor which makes this solar cell configuration one of the simplest photovoltaic cell structures.

Cross-sectional HR-SEM images are shown in Fig. 3.11 for the four different perovskite thicknesses made by 4, 6, 8, and 10 spray passes. Figure 3.11 inset shows that the spray deposition technique creates micron-sized grains of perovskite depending on the number of spray passes (discussed in further detail), minimizing the grain boundaries, which is beneficial for the PV performance.

To investigate the PV performance of these cells, the number of blocking layers, i.e., the thickness of the planar TiO_2 films, was changed and the number of passes were varied. Table 3.1 shows the thickness of the planar TiO_2 film measured by profilometer. As the number of blocking layers increases, the thickness of the planar TiO_2 also increases. The PV results of cells made on different blocking layers with a constant number of passes (10 passes) are presented in Table 3.1 and Fig. 3.12a.

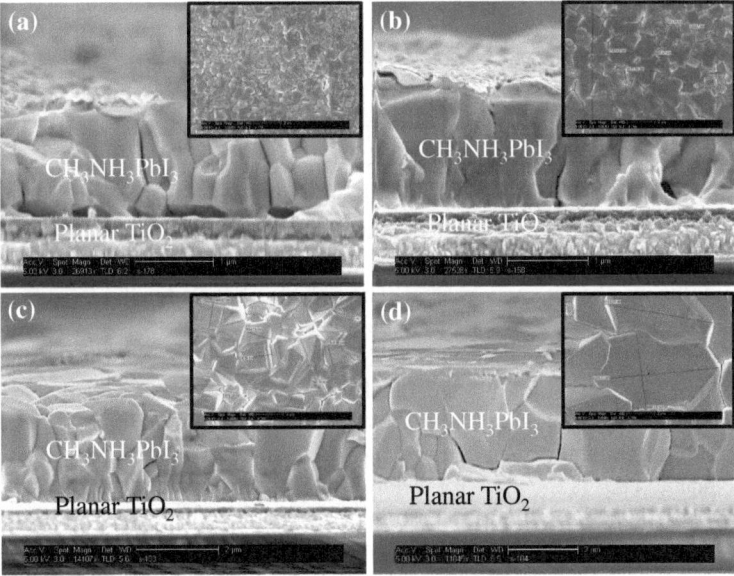

Fig. 3.11 a 4 passes of perovskite spray; **b** 6 passes of perovskite spray; **c** 8 passes of perovskite spray; **d** 10 passes of perovskite spray. Inset—Top view of different spray passes 4, 6 ,8 and 10 respectively. Scale bar at the insets is 2 μm. Reprinted with permission from Ref. [14] Copyright (2015) American Chemical Society

Table 3.1 PV parameters of cells according to various thicknesses of dense TiO$_2$ layers (blocking layers); thicknesses as listed in the table

No. of BL	V_{oc} (V)	J_{sc} (mA/cm^2)	FF (%)	Efficiency (%)	BL thickness (nm)
1	0.55	9.35	39	1.99	74.4 ± 11
2	0.66	21.54	32.7	4.63	76.2 ± 15
3	0.69	23.01	43.4	6.93	166.9 ± 19
4	0.65	21.64	39.7	5.58	184.2 ± 13
5	0.68	17.82	42.3	5.11	227.5 ± 42

The number of spray passes is constant, equal to 10; *BL*—blocking layer

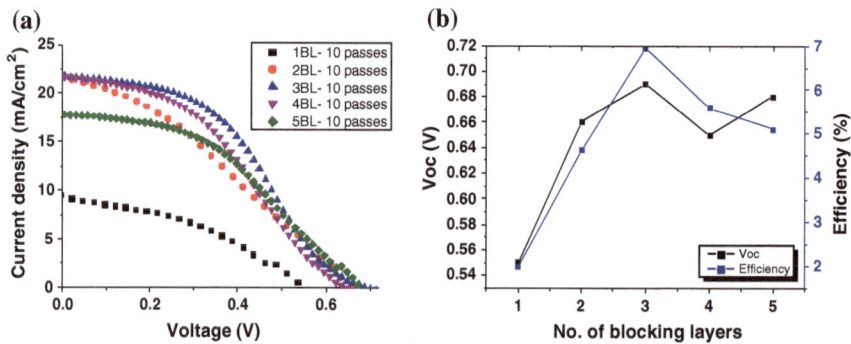

Fig. 3.12 a JV curves of cells made by 10 passes deposited on various blocking layers. **b** Open circuit voltage and efficiency as a function of the number of blocking layers; BL—blocking layer. Reprinted with permission from Ref. [14] Copyright (2015) American Chemical Society

The best PV performance was achieved at 3 blocking layers with open circuit voltage (V_{oc}) of 0.69 V, J_{sc} of 23.01 mA/cm^2 and power conversion efficiency (PCE) of 6.9 % (Fig. 3.12b). It can be observed that the V_{oc} barely changed with the change in the blocking layer thickness, (excluding 1 blocking layer, discussed below) which suggests that the main influence on the V_{oc} is the perovskite film thickness (the number of spray passes). The s-shape observed in the JV curves of all cells might be related to charge accumulation.

The planar HTM free solar cells with 3 blocking layers showed the best PV performance. In the case of 1 and 2 blocking layers, the planar TiO$_2$ was too thin and some holes in the blocking layer were observed, which might contribute to the low performance. On the other hand, in the case of 4 and 5 blocking layers, it seems that the planar TiO$_2$ is too thick for electrons to transport efficiently (also calculated below by the depletion region); therefore, the efficiency decreased.

To optimize the PV performance, the number of passes was changed (while using 3 blocking layers as the planar TiO$_2$) as shown in Fig. 3.13a. It can be seen that the efficiency is the highest in the case of 10 spray passes as shown in Fig. 3.13b. The reason for the best PV performance in the case of 10 spray passes is mainly due to the difference in the perovskite crystal size. As indicated by the

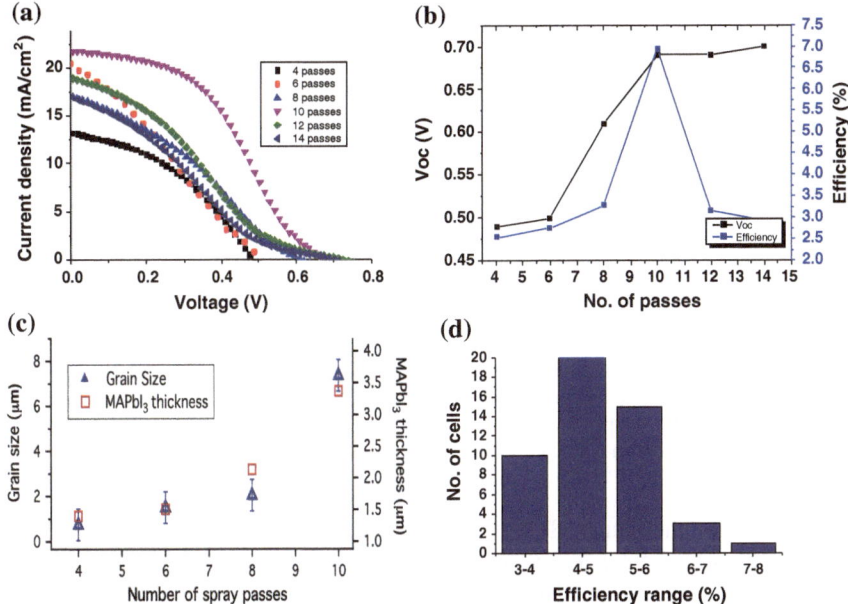

Fig. 3.13 a JV curves of the cells made on 3 blocking layers according to the number of passes. **b** Open circuit voltage and efficiency as a function of the number of passes. **c** Average grain size and the MAPbI$_3$ film thickness as a function of the number of spray passes. **d** Statistics of the HTM-free planar solar cells. Reprinted with permission from Ref. [14] Copyright (2015) American Chemical Society

HR-SEM images in Fig. 3.13, the perovskite crystal size increases with the number of spray passes. In the rest of the discussion we excluded 12 and 14 spray passes since the PV performance decreased significantly in these cases (PCE of 3.1 and 2.9 % of 12 and 14 spray passes respectively). The reason for the decrease in performance for the 12 and 14 passes is due to the fact that the perovskite layer becomes very thick and as a result it is peeled from the surface, which harms the PV performance. Figure 3.13c shows the change in the perovskite crystal size as a function of the spray passes. In the case of 10 spray passes the crystal size of the perovskite is 7.5 ± 1 μm (though, a few crystals were smaller than 6.5 μm and some were larger than 8.5 μm). Nie et al. [15] have reported that large perovskite grains have less defects and higher mobility enabling better charge carrier transport through the perovskite film. Figure 3.13a–c support this argument; the V_{oc} and the efficiency increase with the number of spray passes (bigger perovskite crystals). The increase of the number of spray passes points out the decrease in recombination and defects in the perovskite crystals. Moreover the coverage improved when increasing the number of spray passes further support the increase in the efficiency. Different scan direction and different scan rate in the case of 3 blocking layers with 10 spray passes (the conditions which present the best PV performance) show that the hysteresis is minimal in these cells and almost independent on the scan rate.

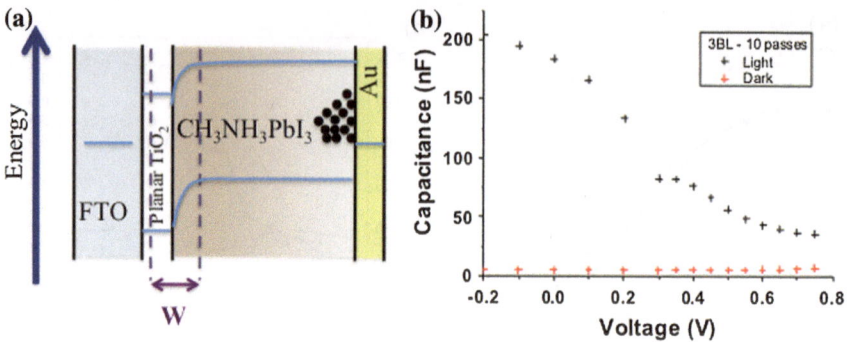

Fig. 3.14 **a** Schematic illustration of the energy level diagram of the planar HTM free perovskite solar cell. **b** Capacitance voltage measurements in the dark and under illumination of the cell with 3 blocking layers and 10 perovskite spray passes. Reprinted with permission from Ref. [14] Copyright (2015) American Chemical Society

Statistics for the planar HTM-free solar cells are presented in Fig. 3.13d. An average efficiency of 4.7 ± 1 % was observed for over 50 cells.

To elucidate the mechanism of these planar HTM-free perovskite solar cells, capacitance voltage measurements were performed both under dark and illumination. Figure 3.14b shows the capacitance voltage measurements for the cell with 3 blocking layers and 10 spray passes of perovskite (this cell demonstrated the best PV performance). All other cells discussed in this manuscript demonstrated the same capacitance voltage behavior as the cell shown in Fig. 3.14b. In the dark, the capacitance remained constant at approximately 5×10^{-9} F. The capacitance under illumination is two orders of magnitude larger than the capacitance in the dark. Under illumination, the capacitance started to increase to several orders of magnitude higher than in the dark, approaching the V_{oc} of the cell, implying that photogenerated carriers are accumulated within the solar cell and cannot be effectively collected by the electrode as shown schematically in Fig. 3.14a. (On the other hand when the capacitance of a solar cell is almost not changing under illumination, photogenerated carriers can be collected effectively by the electrode without charge accumulation.) This observation was also reported by Bisquert et al. [16] where the increase of the capacitance was measured as a function of the applied voltage for nanostructure cells and planar cells, this was related to charge accumulation in the perovskite which indicates a high density of states (DOS) in the perovskite. Since the perovskite thickness in this planar structure is in the range of 1.4–3.4 μm, which is much larger than the perovskite optimum thickness according to its absorption coefficient [17], carriers are accumulated at the interface with the metal contact (supported by the capacitance voltage measurements). The accumulation of carriers might be the explanation to the s-shape observed in the JV curves of these cells. The s-shaped curves contribute to the low fill factor of the planar HTM-free solar cells. It is important to indicate that the observed s-shape was independent on the JV scan rate.

From the discussion presented, it could be construed that thinner perovskite film would be beneficial for the planar HTM-free cell. According to the Fig. 3.13a, b the PCE at 4 passes is much lower than the PCE at 10 passes (which produced thicker and larger perovskite crystals as mentioned). Therefore, it could be argued that besides the charge accumulation at the perovskite/metal interface, there is an additional contribution to the operation mechanism of the planar HTM-free cell.

As described in the previous section, Mott Schottky analysis can be preformed to elucidate the blocking layer (planar TiO_2)/perovskite interface and to extract the depletion region. The results showed similar width of the total depletion region for the three cases (3, 4, and 5 blocking layers). The depletion region width for the 3 BL is slightly wider than the others; however, the results for the perovskite films (Wp = 117–127 nm) indicate that most of the perovskite film is not depleted. At illumination close to the junction the carriers—the holes in this case—are transported to the back contact and cross the whole perovskite film. It is possible that the long diffusion length of electrons and holes is not solely responsible for the operation of this planar HTM-free cell. It is suggested that once a charge separation occurs close to the depletion region, electrons drift to the edge of the depletion region (close to the FTO) while holes are transported to the back contact. In the case of thick perovskite film, the holes are transported without any interference through the perovskite film because far away from the depletion region, there are no free electrons and holes that can recombine with the transported holes. On the other hand, if the charge separation occurs far from the planar TiO_2/perovskite junction, a recombination will probably occur, one of the reasons for the low open circuit voltage observed in these cells.

References

1. Etgar L, Peng G, Xue Z, Liu B, Nazeeruddin MK, Grätzel M (2012) Mesoscopic CH3NH3PbI3/TiO2 heterojunction solar cell. J Am Chem Soc 134:17396–17399
2. Xing G, Mathews N, Sun S, Lim SS, Lam YM, Grätzel M, Mhaisalkar S, Sum TC (2013) Long-range balanced electron- and hole-transport lengths in organic-inorganic CH3NH3PbI3. Science 342:344–347
3. Stranks SD, Eperon GE, Grancini G, Menelaou C, Alcocer MJ, Leijtens T, Herz LM, Petrozza A, Snaith HJ (2013) Electron-hole diffusion lengths exceeding 1 micrometer in an organometal trihalide perovskite absorber. Science 342:341–344
4. Laben WA, Etgar L (2013) Depleted hole conductor-free lead halide iodide heterojunction solar cell. Energy Environ Sci 6:3249–3253
5. Aharon S, Gamliel S, Cohen B, Etgar L (2014) Depletion region effect of highly efficient hole conductor free CH3NH3PbI3 perovskite solar cells. Phys Chem Chem Phys 16:10512–10518
6. Cohen Bat-El, Aharon Sigalit, Dymshits Alexander, Etgar L (2015) Impact of anti-solvent treatment on carrier density in efficient hole conductor free perovskite based solar cells. J Phys Chem C. doi:10.1021/acs.jpcc.5b10994
7. Cohen BE, Gamliel S, Etgar L (2014) Parameters influencing the deposition of methylammonium lead halide iodide in hole conductor free perovskite-based solar cells. APL Mater 2(081502)

8. Burschka J, Pellet N, Moon S-J, Humphry-Baker R, Gao P, Nazeeruddin MK, Graïzel M (2013) Sequential deposition as a route to high-performance perovskite-sensitized solar cells. Nature 499:316–319
9. Luber JE, Buriak MJ (2013) Reporting performance in organic photovoltaic devices. ACS Nano 7:4708–4714
10. Mashkoor A, Jiong Z, Javed I, Wei M, Lin X, Rigen M, Jing Z, Mashkoor A et al (2009) Conductivity enhancement by slight indium doping in ZnO nanowires for optoelectronic applications. Phys D Appl Phys 42:165406
11. Strukov1 DB, Snider1 GS, Stewart1 DR, Williams S (2008) The missing memristor found. Nature 453:80–83
12. Jeon NJ, Noh HJ, Kim CY, Yang SW, Ryu S, SeokNam S (2014) Solvent engineering for high-performance inorganic-organic hybrid perovskite solar cells. Nat Mater 13:897–903
13. Tripathi N, Yanagida M, Shirai Y, Masuda T, Hanb L, Miyanoa K (2015) Hysteresis-free and highly stable perovskite solar cells produced via a chlorine-mediated interdiffusion method. J Mater Chem A 3:12081–12088
14. Gamliel S, Dymshits A, Aharon S, Terkieltaub E, Etgar L (2015) Micrometer sized perovskite crystals in planar hole conductor free solar cells. J Phys Chem C. doi:10.1021/acs.jpcc.5b07554
15. Nie W, Tsai H, Asadpour R, Blancon JC, Neukirch AJ, Gupta G, Crochet JJ, Chhowalla M, Tretiak S, Alam MA, Wang HL, Mohite AD (2015) High-efficiency solution-processed perovskite solar cells with millimeter-scale grains. Science 347:522–525
16. Kim H-S, Mora-Sero I, Pedro VG, Santiago FF, Juarez-Perez EJ, Park N-G, Bisquert J (2013) Mechanism of carrier accumulation in perovskite thin-absorber solar cells. Nat Commun 4:2242
17. De Wolf S, Holovsky J, Moon SJ, Löper P, Niesen B, Ledinsky M, Haug FJ, Yum JH, Ballif C (2014) Organometallic halide perovskites: sharp optical absorption edge and its relation to photovoltaic performance. J Phys Chem Lett 5:1035–1039. doi:10.1021/jz500279b
18. Cohen Bat-El, Aharon Sigalit, Dymshits Alexander, Etgar L (2015) Impact of anti-solvent treatment on carrier density in efficient hole conductor free perovskite based solar cells. J Phys Chem C. doi:10.1021/acs.jpcc.5b10994

Chapter 4
Parameters Influencing the Deposition of Methylammonium Lead Halide Iodide in Hole Conductor Free Perovskite-Based Solar Cells

4.1 Introduction

Several deposition techniques are used for perovskite-based solar cells, including vapor deposition, vapor assisted solution process (VASP), and solution processed via one-step and two-step deposition [1–5]. The effect of the perovskite deposition on the solar cells performance is critically important; it determines the film coverage, film thickness, film quality, and the transport properties. Graetzel and coworkers have demonstrated the use of the two-step deposition technique as a powerful technique for achieving highly efficient perovskite solar cells. The two-step deposition enables better control over the perovskite crystallization by separating the perovskite deposition into two precursors.

This chapter discusses the study of different parameters influencing the two-step deposition of $CH_3NH_3PbI_3$ perovskite in photovoltaic solar cells. The perovskite solar cell structure discussed is hole conductor free architecture as presented earlier.

Two-step deposition includes several stages: (a) dropping the PbI_2 solution onto the TiO_2 electrode; (b) spin coating the PbI_2 and annealing the PbI_2. (c) dipping the PbI_2 electrode into CH_3NH_3I (CH_3NH_3I = MAI) solution; (d) annealing of the $CH_3NH_3PbI_3$ perovskite. Figure 4.1 shows the different stages in the deposition process which will be discussed below.

4.2 Spin Velocity

The influence of spin velocity on PV performance was examined by changing the spin velocity of the PbI_2 on the TiO_2 film (process 1 in Fig. 4.1).

Figure 4.2 presents a top view HR-SEM of the $CH_3NH_3PbI_3$ at different spin velocities. The dipping time (related to process 2 in Fig. 4.1, discussed in Sect. 4.3) was the same for all spin velocities. It is seen that for 500 rpm the coverage of the

© The Author(s) 2016
L. Etgar, *Hole Conductor Free Perovskite-based Solar Cells*,
SpringerBriefs in Applied Sciences and Technology,
DOI 10.1007/978-3-319-32991-8_4

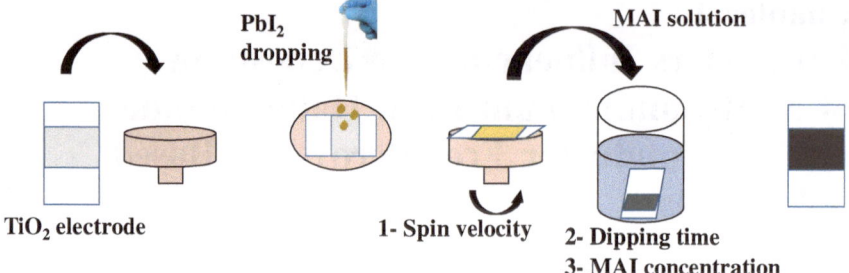

Fig. 4.1 The different stages involved in the two-step deposition of the CH$_3$NH$_3$PbI$_3$ perovskite. Methylammonium iodide (MAI) corresponds to CH$_3$NH$_3$I. Taken with permission from Applied Physics letter Materials Ref. [5]

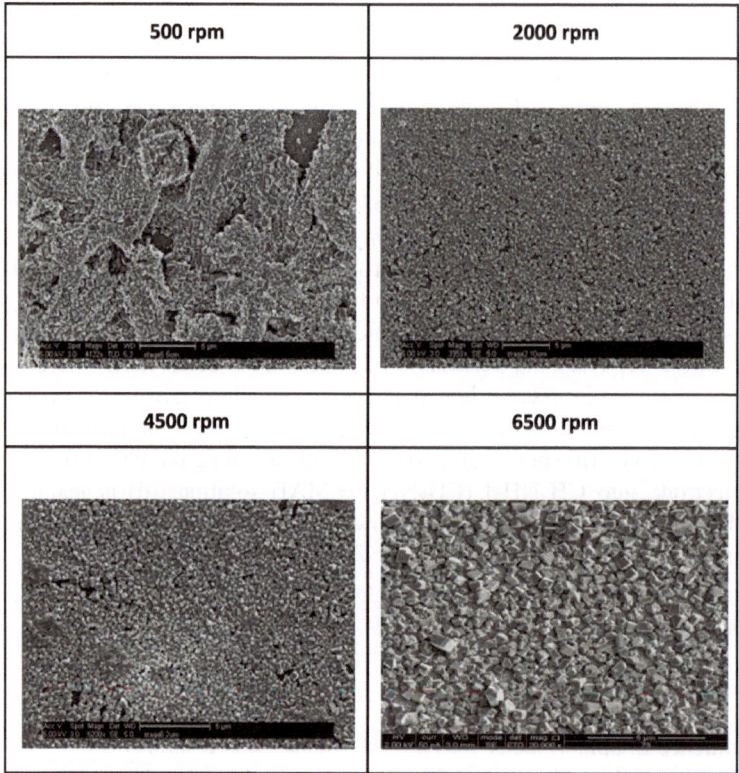

Fig. 4.2 Top view HR-SEM micrographs of CH$_3$NH$_3$PbI$_3$ crystals deposited on TiO$_2$ film at different spin velocities (rpm–round per minute) Taken with permission from Applied Physics letter Materials Ref. [5]

Table 4.1 PV parameters of the cells at different spin velocities

Spin velocity (rpm)	V_{oc} (V)	J_{sc} (mA/cm^2)	FF (%)	Efficiency (%)
500	0.82	15.3	45	5.6
2000	0.80	19.1	51	7.8
4500	0.86	17.8	47	7.2
6500	0.81	15.2	56	6.9

$CH_3NH_3PbI_3$ is not complete; there are several voids. When the spin velocity is increased, the PbI_2 film is more uniform than when the spin velocity is low (low rpm) as seen at 2000, 4500, and 6500 rpm. The $CH_3NH_3PbI_3$ crystals increase in size with increased spin velocity. The reason for that is related to the dipping time in the CH_3NH_3I solution (process 2 in Fig. 4.1). As indicated, the dipping time in this set of experiments was the same—at 2000 rpm, the PbI_2 film is thicker than at 4500 rpm, while the thinnest PbI_2 film was observed at 6500 rpm. Different film thicknesses require different dipping times. However, when the dipping time is the same, the resulting crystals' size is dependent on the spin velocity (film thickness); therefore, in the case of 6500 rpm (the thinnest PbI_2 film thickness), the perovskite crystals were the biggest.

Table 4.1 summarizes the PV parameters of the cells at different spin velocities. The best PV results are achieved for the 2000 rpm sample with power conversion efficiency (PCE) of 7.8 %. The data shows that by changing the spin velocity, the current density of the solar cell is the parameter with the most influence.

4.3 Dipping Time

The effect of dipping time in the CH_3NH_3I solution (process 2 in Fig. 4.1) on morphology and on PV performance was studied. Figure 4.3 presents dipping times of 20 s, 1 min, 20 min, and 3.5 h. It is noted that increasing the dipping time increases the size of the $CH_3NH_3PbI_3$ perovskite crystals (spin velocity was the same for all samples in these experiments), resulting in perovskite crystals the size of several micrometers.

The influence of dipping on PV performance is observed in Table 4.2 and Fig. 4.4. The best PV performance was observed for the shortest dipping time of 20 s. Longer dipping time results in bigger crystals which affect transport through the perovskite film and enhance the recombination at longer dipping times, as indicated by the dark current measurements in Fig. 4.4a. The V_{OC} and FF were lower for dipping times of 3.5 h and for 1 h, compared to the other dipping times.

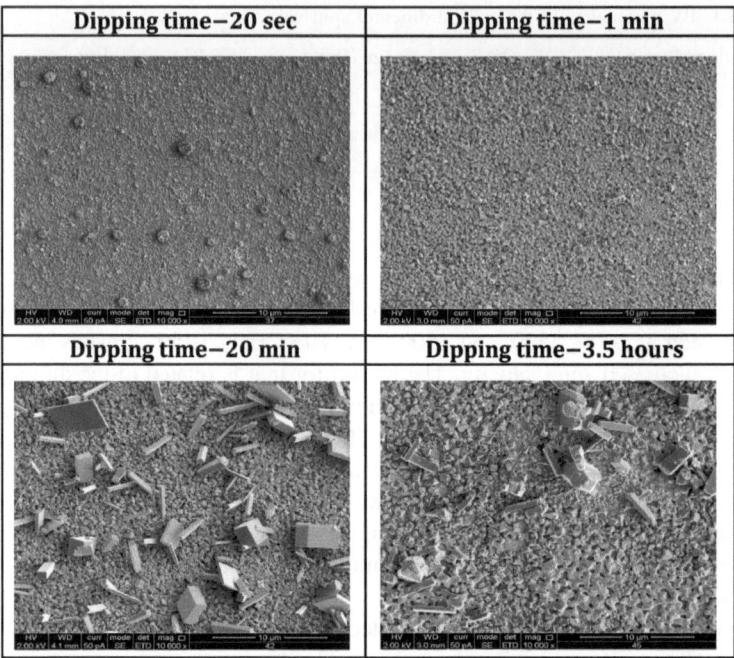

Fig. 4.3 Top view HR-SEM micrographs of CH₃NH₃PbI₃ crystals deposited on TiO₂ film at four different dipping times. Taken with permission from Applied Physics letter Materials Ref. [5]

Table 4.2 PV performance of the cells at 5 different dipping times

Dipping time	V_{oc} (V)	J_{sc} (mA/cm^2)	FF (%)	Efficiency (%)
20 s	0.77	12.1	66	6.2
1 min	0.75	9.4	61	4.4
20 min	0.72	7.4	67	3.6
1 h	0.58	0.9	43	0.3
3.5 h	0.22	1.0	37	0.1

4.4   Annealing Temperature

The annealing step is the final stage before evaporation of the contact. (As indicated previously, no hole transport material is used in these cells). To investigate the effect of the annealing temperature, several annealing temperatures were studied. Figure 4.5 presents HR-SEM micrographs of four annealing temperatures—100, 150, 170, and 200 °C. At 100 °C annealing temperature, the CH₃NH₃PbI₃ perovskite crystals are separated from each other. When increasing the annealing temperature to 150 °C, some of the perovskite crystals are sintered together, while at 170 °C annealing temperature, most of the perovskite crystals are sintered and

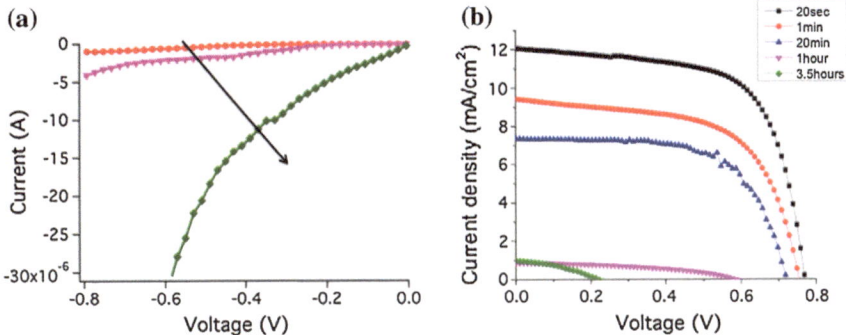

Fig. 4.4 **a** Dark current for various dipping times: 1 min, 1 h, 3.5 h. The *arrow* indicates the increase in the dark current. **b** Current voltage curves of the hole conductor free solar cells at different dipping times. Taken with permission from Applied Physics letter Materials Ref. [5]

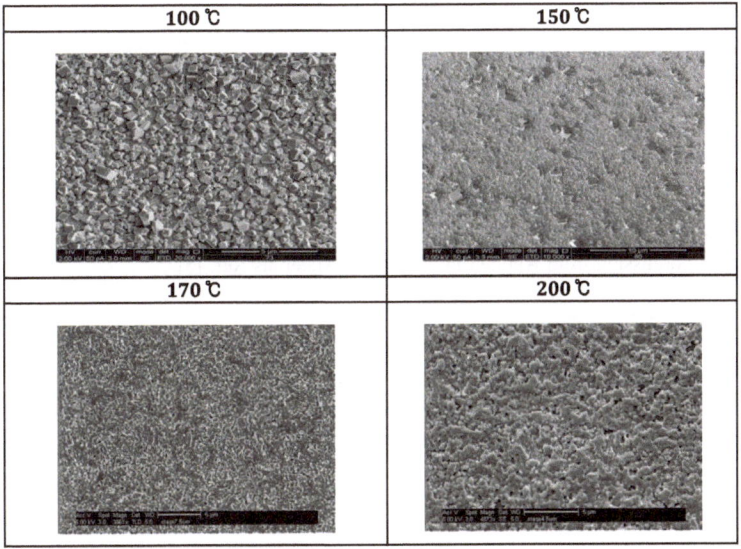

Fig. 4.5 Top view HR-SEM micrographs of $CH_3NH_3PbI_3$ crystals deposited on TiO_2 film and annealed at different temperatures. Taken with permission from Applied Physics letter Materials Ref. [5]

there are no separate crystals of perovskite at the surface. At 200 °C annealing temperature, the $CH_3NH_3PbI_3$ perovskite crystals are melted, resulting in the lowest power conversion efficiency.

The PV performance and the current voltage curves of the different annealing temperatures are presented in Table 4.3 and Fig. 4.6b. The best photovoltaic performance was achieved at an annealing temperature of 170 °C, and the highest

Table 4.3 PV results of different annealing temperatures

Annealing Temperature (°C)	V_{oc} (V)	J_{sc} (mA/cm²)	FF (%)	Efficiency (%)
70	0.81	10.7	61	5.4
100	0.86	14.5	55	6.8
150	0.77	13.0	63	6.1
170	0.80	17.7	62	8.7
200	0.70	5.4	50	1.9

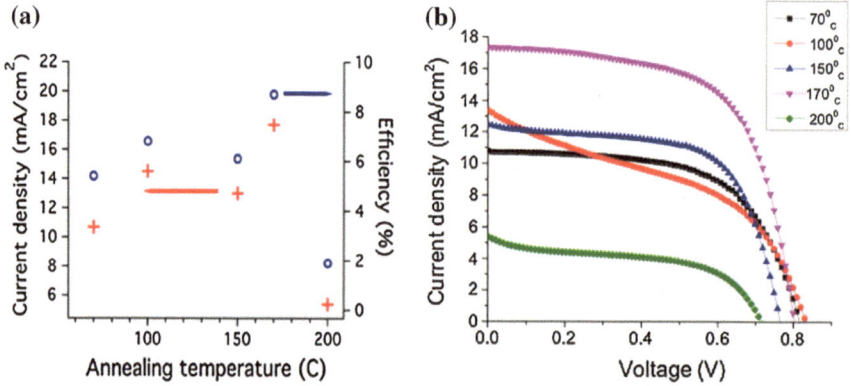

Fig. 4.6 a Current density and efficiency as a function of the annealing temperature. Current voltage curves of the hole conductor free perovskite solar cells annealed at different temperatures. Taken with permission from Applied Physics letter Materials Ref. [5]

current density was observed at this annealing temperature, probably a result of better sintering of the $CH_3NH_3PbI_3$ perovskite crystals than at the lower temperatures (70, 100, and 150 °C). Figure 4.6a summarizes the influence of the annealing temperature on the current density and the efficiency of the hole conductor free solar cells. Obviously, the annealing temperature mainly influences the current density.

4.5 Methylammonium Iodide Concentration

The concentration of MAI in the dipping solution has an important effect on the cells performance (process 3 in Fig. 4.1). Increasing the MAI concentration results in more MAI to react with the already deposited PbI_2, which could increase the crystallization of the $CH_3NH_3PbI_3$ perovskite. However, when an excess of MAI is present in the dipping solution, the free MAI (which did not react with the PbI_2) could have the opposite effect. The excess MAI might remain on the $CH_3NH_3PbI_3$ perovskite surface, reducing the perovskite conductivity. Moreover, excess MAI could cause desorption of the perovskite from the surface. Table 4.4 shows the PV

Table 4.4 PV parameters of cells made at different MAI concentrations in the dipping solution

MAI concentration [M]	V_{oc} [V]	J_{sc} [mA/cm^2]	FF (%)	Efficiency (%)
0.06	0.80	19.1	51	7.8
0.09	0.77	18.9	65	9.4
0.12	0.84	16.9	47	6.6
0.15	0.8	10.9	58	4.8

Fig. 4.7 Fill factor and efficiency of the hole conductor free perovskite solar cells at different MAI concentrations. Taken with permission from Applied Physics letter Materials Ref. [5]

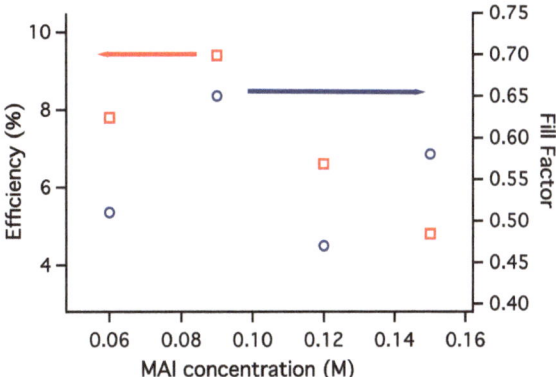

parameters at different MAI concentrations in the dipping solution. At high MAI concentration, the PV performance decreases due to a decrease in the current density. For low MAI concentration, the current density is high, although the FF is lower than the 0.09 M MAI. Table 4.4 shows that there is an optimum concentration where the FF and the current density are the highest, resulting in PCE of 9.4 %, the highest efficiency in this study. Above this optimum point, both fill factor and efficiency decreased due to an excess of MAI in the dipping solution (Fig. 4.7).

In this chapter, the two-step deposition technique was used for the deposition of the CH$_3$NH$_3$PbI$_3$ in hole conductor free perovskite solar cells. The elimination of hole conductor in this study enabled isolating the effect of different parameters in the two-step deposition process and to investigate their influence on the PV performance.

The effect on the photovoltaic performance was investigated by changing several parameters in the two-step deposition, i.e., spin velocity, dipping time, annealing temperature, and various methylammonium iodide concentrations. It was concluded that the spin velocity has the most influence on the J_{SC}, dipping time on the FF, annealing temperature on the J_{SC}, and MAI concentration on the FF and J_{SC}. Interestingly, the V_{OC} was almost not affected by these parameters.

Understanding the effect of these critical parameters on perovskite deposition could lead to highly efficient, low-cost perovskite-based solar cells.

References

1. Chen Q, Zhou H, Hong Z, Luo S, Duan H-S, Wang H-H, Liu Y, Li G, Yang Y (2014) Planar heterojunction perovskite solar cells via vapor-assisted solution process. J Am Chem Soc 136:622–625
2. Liu M, Johnston MB, Snaith HJ (2013) Efficient planar heterojunction perovskite solar cells by vapour deposition. Nature 501:395–399
3. Burschka J, Pellet N, Moon S-J, Humphry-Baker R, Gao P, Mohammad NK, Graetzel M (2013) Nature 499:316
4. Kitazawa N (1997) Excitons in two-dimensional layered perovskite compounds: $(C_6H_5C_2H_4NH_3)_2Pb(Br, I)_4$ and $(C_6H_5C_2H_4NH_3)_2Pb(Cl, Br)_4$. Mater Sci Eng B 49:233–238
5. Cohen BE, Gamliel S, Etgar L (2014) Parameters influencing the deposition of methylammonium lead halide iodide in hole conductor free perovskite-based solar cells. APL Mater 2(081502)

Chapter 5
Tuning the Optical Properties of Perovskite in HTM Free Solar Cells

As discussed in the introduction of this book, the optical properties of OMHP could be changed by chemical modifications, in the halide site (X-site) or in the cation site (A-site), see Fig. 1.1. This chapter discusses both options and their function in HTM-free OMHP solar cell.

5.1 Halide (X-Site) Modifications [1]

The most common halides in the X-site are Cl, I, and Br. Br and I halides are most influencing the optical properties of the perovskite. While exchanging all the halides by Cl, the OMHP has no color which suggests that a wide band gap material was formed.

Figure 5.1a shows the structure of the hole-conductor-free solar cell. The solar cell consists of conductive glass, and thin TiO_2 nanoparticles film; on top of the TiO_2 film, a perovskite layer is deposited which consists of perovskite composition according to the formula $CH_3NH_3PbI_nBr_{3-n}$ (where $0 \leq n \leq 3$). No hole conductor is used in these solar cells which result in hole conductivity through the hybrid $CH_3NH_3PbI_nBr_{3-n}$. The energy level diagram is presented in Fig. 5.1b; both pure $CH_3NH_3PbI_3$ and $CH_3NH_3PbBr_3$ can inject electrons to the TiO_2 and transfer the holes to the gold back contact; as a result, the energy level position of the hybrid structure $CH_3NH_3PbI_nBr_{3-n}$ would be suitable for the operation of the cell.

The deposition of the Hybrid $CH_3NH_3PbI_nBr_{3-n}$ (where $0 \leq n \leq 3$) perovskites was done by a two-step deposition technique, which allows better control of the hybrid perovskite composition and its corresponding band gap.

The thickness of the TiO_2 film was 500 ± 50 nm (measured by profiler) and the perovskite film was 150 ± 50 nm thick. Possibly some of the perovskite was penetrated inside the pores; this results with an efficient electron transfer between the perovskite and the TiO_2 [2]. On the other hand we can assume that the recombination occurred in these cells is mainly between electrons in the TiO_2 and

© The Author(s) 2016
L. Etgar, *Hole Conductor Free Perovskite-based Solar Cells*,
SpringerBriefs in Applied Sciences and Technology,
DOI 10.1007/978-3-319-32991-8_5

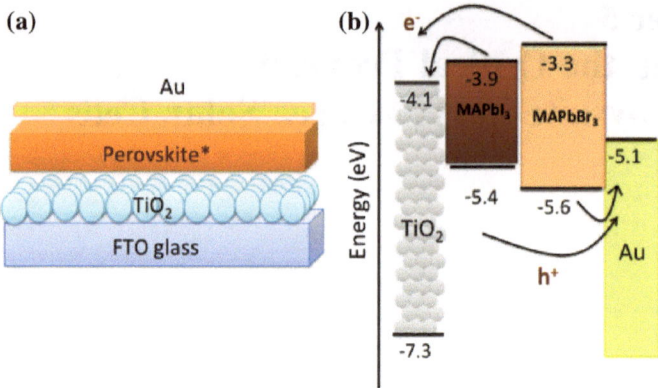

Fig. 5.1 a Structure of the hybrid lead halide iodide and lead halide bromide hole-conductor-free perovskite solar cell. The *perovskite** indicates that the perovskite structure consists of different MAI and MABr (CH₃NH₃ = MA) concentrations investigated in this study. **b** Energy level diagram of the pure MAPbI₃ and the pure MAPbBr₃ perovskites used in this work. The positions of the energy levels are according to [17]. Reprinted with permission from Ref. [1] Copyright (2014) American Chemical Society

holes in the perovskite, therefore we expect that less recombination will occur when the pores are not completely filled. In this work high V_{oc} of around 0.8 V and high FF of 68 % were achieved indicating that the pores probably are not completely filled with perovskite (these are relatively high values when discussing on the situation where no hole conductor is being used).

When comparing two different dipping time in the cation solution (MAI or MABr) a change in the morphology of the crystals can be observed (Fig. 5.2). In the case of the PbI₂ + MAI (CH₃NH₃PbI₃) no difference is observed between 2 s of dipping and 30 s of dipping. However, when looking on different dipping time in the case of PbI₂ + MABr, after 2 s of dipping there are still areas where there are not any perovskite crystals while after 30 s of dipping most the crystals are formed. This suggests that the cation interacts differently depends on its halide.

Changing the halogen is affecting the optical properties of the perovskite. Figure 5.3 presents the absorption spectra of the various perovskite compositions, the PbBr₂ + MABr:MAI 1:0 indicates the CH₃NH₃PbBr₃ perovskite with a band gap of 2.12 eV, while the PbI₂ + MABr:MAI 0:1 indicates the CH₃NH₃PbI₃ perovskite with a band gap of 1.51 eV. The other samples show different concentrations of MAI and MABr in the dipping solution corresponding to various band gaps. The range of the bang gaps, are between 1.51 and 2.12 eV, suitable for the hole-conductor-free solar cells since the conduction band position is higher than the TiO₂ conduction band, and the valence band position is lower than the gold work function.

As described above, the exciton absorption bands shifted to short wavelength (higher photon energy) when introducing the Br ions. The position of the absorption peak is determined by the Br(4p) orbitals with the Pb(6s) orbitals (related to the valance band of the CH₃NH₃PbI$_n$Br$_{3-n}$ perovskite). The transitions in the

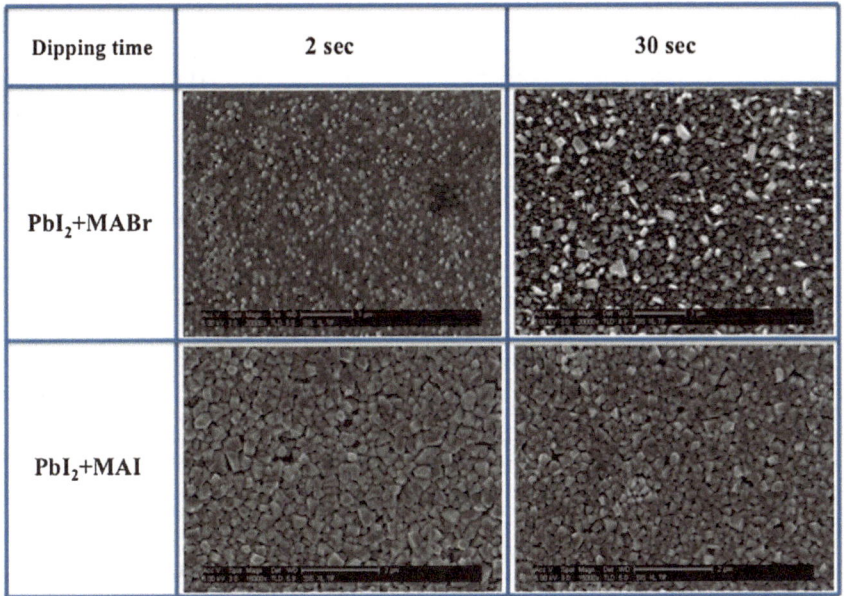

Dipping time	2 sec	30 sec
PbI₂+MABr		
PbI₂+MAI		

Fig. 5.2 HR-SEM figures of PbI_2 + MABr and PbI_2 + MAI samples at different dipping time in the MABr or MAI solution, respectively. Reprinted with permission from Ref. [1] Copyright (2014) American Chemical Society

Fig. 5.3 Absorption spectra of the different perovskites composition. The *arrow* indicates the increase in *n*. Reprinted with permission from Ref. [1] Copyright (2014) American Chemical Society

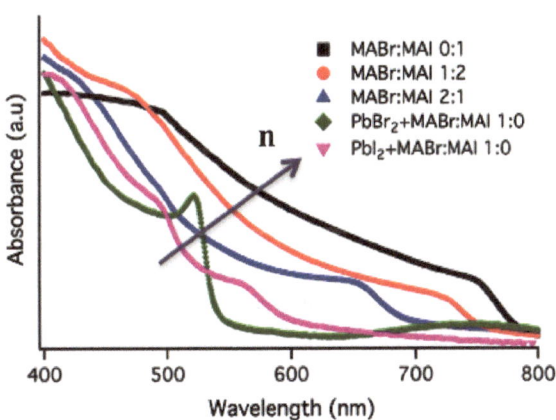

methylammonium lead halide perovskite are similar to the transitions in PbI_2 [3, 4]. The valance band of the PbI_2 composed of Pb(6s) orbitals and I(5p) orbitals while the conduction band composed of Pb(6p) orbitals. Moreover, the energy level of Br (4p) is lower than the energy level of Pb(6s), therefore the peak position of $CH_3NH_3PbI_nBr_{3-n}$ is influenced and shifted to higher energy [5].

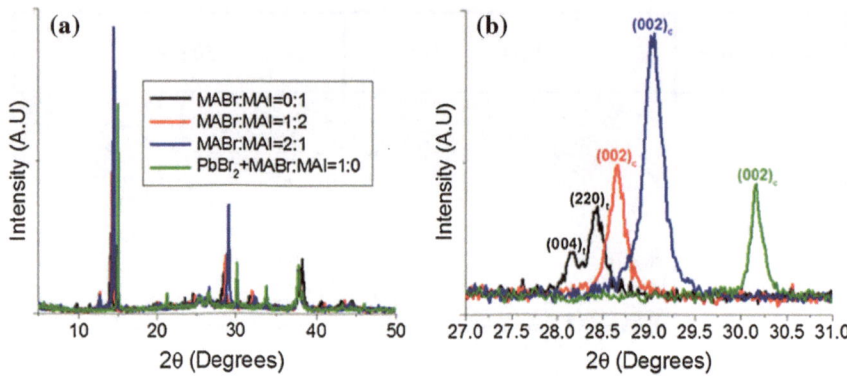

Fig. 5.4 a XRD diffraction of the different samples contains different Br⁻ ions concentrations. **b** XRD spectra in the range of $27° \leq 2\theta \leq 31°$. The miller indexes are mentioned on the figure. 'c' corresponds to cubic and 't' corresponds to tetragonal. Reprinted with permission from Ref. [1] (2014) American Chemical Society

The XRD spectra are shown in Fig. 5.4a. The diffraction peaks of the various cells are observed, showing the crystallographic structure of the samples. In all cells where Br⁻ ions are involved, the diffraction peaks correspond to the cubic structure, although in the case of $CH_3NH_3PbI_3$ (where there are no Br⁻ ions), the peaks are related to the tetragonal structure. Figure 5.4b shows a magnification of the XRD spectra in the range of $27° \leq 2\theta \leq 31°$.

The position of the (002) diffraction peak for the samples is changed according to the Br⁻ concentrations. In the case of the $CH_3NH_3PbI_3$, two peaks can be observed, indexed as (220) and (004) related to the tetragonal structure. These two diffraction peaks are merged into one when introducing the Br into the perovskite structure. As a result of the implementation of the Br⁻ ions into the perovskite structure, the lattice parameter changes from 5.921 for the $CH_3NH_3PbBr_3$, 6.144 for the MABr:MAI 2:1 and 6.223 for the MABr:MAI 1:2. The change in the lattice parameter is due to the difference in the ionic radius of Br⁻ (1.96A) and I⁻ (2.2A) [6]. The smaller ionic radius of the Br⁻ is the main reason for the formation of the cubic structure when Br is introduced into the perovskite structure.

The photovoltaic results of the hybrid perovskite hole-conductor-free solar cells are summarized in Table 5.1. An important, novel observation from these results is the possibility of the various perovskites containing Br to conduct holes (since no hole transport material is used in these cells). The pure $CH_3NH_3PbI_3$ achieved power conversion efficiency (PCE) of 7.2 % at a light illumination of 100 mW cm^{-2} AM 1.5G. On the other hand, the $CH_3NH_3PbBr_3$ achieved the lowest PCE of 1.69 %. PCE of 8.54 % was achieved with the molar ratio of MABr to MAI (in the dip solution) as 1:2.

A possible reason for the this PCE could be due to the suitable band gap determined by the iodide, while the partial substitution of the bromide could

Table 5.1 The photovoltaic parameters of the solar cells and their energy band gap

Spin solution	MABr:MAI molar ratio in dip solution	V_{oc} (V)	J_{sc} (mA/cm^2)	FF	Efficiency (%)	E_g (eV)	R_s (Ω)
PbI$_2$	0:1 (CH$_3$NH$_3$PbI$_3$)	0.77	15.6	60	7.2	1.51	65.6
PbI$_2$	1:2	0.77	16.2	68	8.54	1.58	61.7
PbI$_2$	2:1	0.71	13.5	58	5.64	1.65	76.7
PbI$_2$	1:0	0.75	9.6	35	2.55	1.72	185.5
PbBr$_2$	1:0 (CH$_3$NH$_3$PbBr$_3$)	0.79	4.7	46	1.69	2.12	350.3

provide better stability and has smaller effect on the photovoltaic performance, finally the combination of both achieved higher PCE than the CH$_3$NH$_3$PbI$_3$.

5.2 Cation (A-Site) Modifications [7]

Interestingly changing the organic cation to other organic cation or even possibly to inorganic cation could also modify the optical properties of the OMHP and in some cases to enhance its stability.

In this work, we demonstrated that Formamidinium Iodide (FAI) could replace the methylammonium iodide cation in the OMHP structure and could function in a hole-conductor-free (MA/FA)PbI$_3$-based solar cells configuration, where the (MA/FA)PbI$_3$ acts both as a light harvester and as a hole transporter. The two-step deposition was used to fabricate MAPbI$_3$, FAPbI$_3$, and MAPbI$_3$:FAPbI$_3$ = 1:1 perovskites as the light harvester in the solar cell. The effect of composition and annealing temperature on the photovoltaic performance (PV) was studied. Surface photovoltage spectroscopy was used to elucidate the electronic behavior of these perovskites. Intensity modulated photovoltage spectroscopy (IMVS) and intensity modulated photocurrent spectroscopy (IMPS) were used to reveal more details about the diffusion length, charge collection efficiency and recombination lifetime. Moreover, the PV performance of the cells was measured at various temperatures. Analyzing the solar cells performance at different temperatures in the range of 22–95 °C is highly important: First, the organic part of the hybrid perovskite is mobile at this range of temperatures, which could affect the PV performance. Second, in practice the solar cell should be working in this range of temperatures (e.g., on roofs, inside cars, etc.). Therefore, analyzing the PV performance at different temperatures should provide essential information on the cell behavior at working conditions.

The two different perovskites functioned both as a light harvesters and a hole conductors. The structure of the hole-conductor-free perovskite solar cell is composed of FTO glass/TiO$_2$ compact layer/Mesoporous TiO$_2$/Perovskite/gold, as presented in Fig. 5.5a. The perovsksites were deposited by two-step deposition where MAPbI$_3$, FAPbI$_3$, and a mixture of (MA:FA)PbI$_3$ with a ratio of 1:1 were

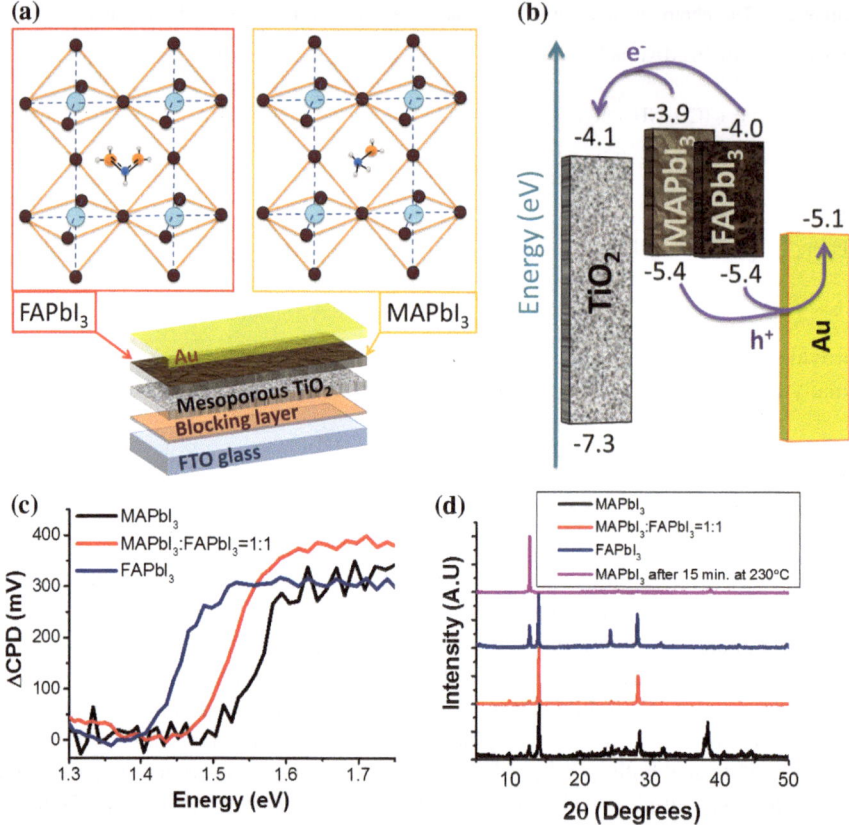

Fig. 5.5 **a** The structure of the cell: FTO glass/blocking layer (compact TiO_2)/mesoporous TiO_2/hybrid organic–inorganic Perovskite/gold. The layer of perovskite contains $MAPbI_3$, $FAPbI_3$, or a mixture of both. The colors in the crystalline structures represent Pb (*light blue*), iodide (*brown*), carbon (*blue*) nitrogen (*orange*) and hydrogen (*white*). **b** Energy levels of the different layers of the cell [17]. **c** SPS measurement of the different perovskites. **d** XRD of the different samples: $MAPbI_3$, $MAPbI_3$: $FAPbI_3$ = 1:1, $FAPbI_3$, and $MAPbI_3$ after 15 min at 230 °C. Taken with permission from Ref. [7]

studied. During illumination, the electrons are injected to the mesoporous TiO_2 from the light harvester material (i.e., the perovskites, which could be $MAPbI_3$, $FAPbI_3$, or their mixture), and the holes are transported to the gold contact. Electron injection and hole transport is possible for the three perovskite compositions.

Changing the cation from FA to MA could tune the band gap of the material. To determine the band gap, optical measurements and Surface Photovoltage Spectroscopy (SPS) was used [8, 9]. The SPV/SPS experiment is based on the classical Kelvin probe technique, which measures the difference in work functions [also known as the contact potential difference (CPD)] between a metallic reference probe and the semiconductor surface. The distance between the reference electrode and semiconductor is a few millimeters while a capacitor arrangement is formed. Once the metallic probe vibrates, an AC capacitance is formed between the probe and the semiconductor results in an AC current in the external circuit. This AC current is zero if and only if there is no charge on the capacitor. In this case, the CPD must be zero. When a DC bias is applied it nullifies the AC current. Thus, the applied DC bias is equal and opposite to the CPD.

In this work, the scanned surfaces were $MAPbI_3$, $FAPbI_3$, and a mixture of both which owned a 1:1 molar ratio. Figure 5.6c shows the ΔCPD versus the energy in eV of the studied samples. The first observation that arises from the SPS spectra in Fig. 5.5c is that $MAPbI_3$ and $FAPbI_3$ (and their mixture) are p-type semiconductors. This is because of the fact that in the case of p-type semiconductors there is a downward band-bending of the conduction band towards the surface at the junction between the metal and the semiconductor (whereas in the case of n-type semiconductors, an upwards band-bending occurs). Once the applied energy is higher

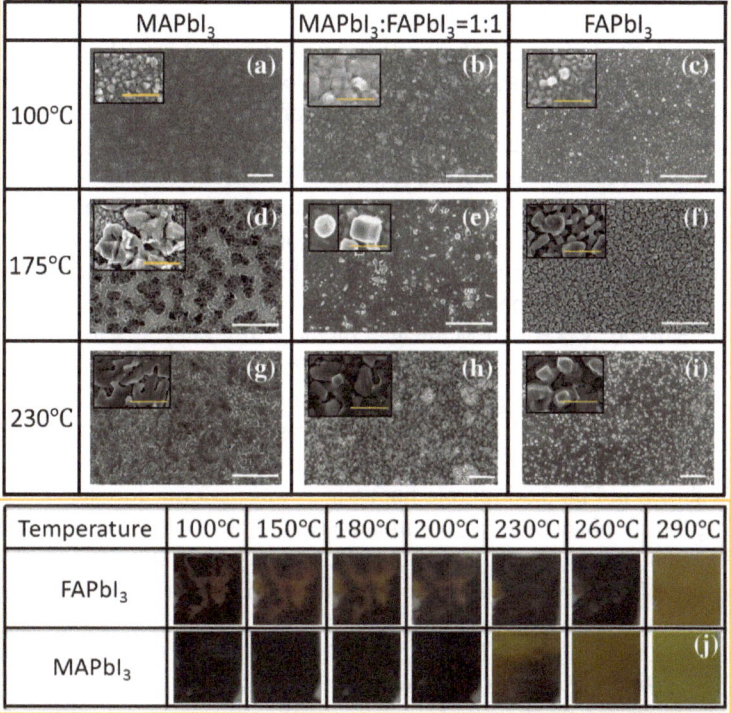

Fig. 5.6 UHR-SEM images of the Perovskite compositions at various annealing temperatures (100, 175, and 230 °C). The *white scale bar* is 5 μm and the *orange scale bar* (*inset*) is 1 μm. **a** $MAPbI_3$ annealed at 100 °C. **b** $MAPbI_3$: $FAPbI_3$ = 1:1 annealed at 100 °C. **c** $FAPbI_3$ annealed at 100 °C. **d** $MAPbI_3$ annealed at 175 °C. **e** $MAPbI_3$: $FAPbI_3$ = 1:1 annealed at 175 °C. **f** $FAPbI_3$ annealed at 175 °C. **g** $MAPbI_3$ annealed at 230 °C. **h** $MAPbI_3$: $FAPbI_3$ = 1:1 annealed at 230 °C. **i** $FAPbI_3$ annealed at 230 °C. **j** Pictures of the samples of $MAPbI_3$ or $FAPbI_3$ on glass after annealing at various temperatures. Taken with permission from Ref. [7]

than the band gap, the bending is decreased, meaning that there is a positive ΔCPD in p-type semiconductors (and negative ΔCPD in the case of n-type semiconductors) [10]. This can be seen by the sign of the knee associated with surface photovoltage onset.

In addition to the electronic behavior of the different perovskites, the band gap energy of each material can be extracted using the SPS spectra. The SPS method is immune to reflection or scattering losses; only photons that are absorbed in the sample contribute to the SPS signal. In this case, the signal onset starts at photon energies close to the band gap of the perovskite samples ($FAPbI_3$, $MAPbI_3$ and the mixture). As a result, it can be concluded that the perovskite samples have fewer sub-band gap states. The band gaps (E_g) of $MAPbI_3$, $FAPbI_3$ and the mixture are estimated to be 1.57, 1.45 and 1.53 eV, respectively (error of ±0.01 eV). It is important to note that the SPS measurements were performed under high level of accuracy.

The band gaps of the perovskites were also estimated using Tauc plots by manipulating transparency measurements of the perovskite films (Not shown here). The E_g values of $MAPbI_3$, $FAPbI_3$ and the mixture obtained by the Tauc plot analysis are 1.56, 1.46 and 1.54 eV, respectively (error of ±0.01 eV, resulted by the linear fitting). There is good agreement between the results of the two methods, and this fact supports the validity of each one of the methods.

Moreover, the results are in good agreement with theory, which predicts that the larger the cation, the smaller the E_g [11]. The XRD spectra of the three perovskites are presented in Fig. 5.5d. The fact that the mixture shows peaks from both samples ($MAPbI_3$ and $FAPbI_3$) indicates that a mixture of the two samples was formed. In addition, the E_g of the mixture is between the band gap energies of $MAPbI_3$ and $FAPbI_3$ which further support this claim.

Figure 5.6 shows XHR-SEM images of the perovskites at various annealing temperatures. Figure 5.6a, d, g show the $MAPbI_3$ at various annealing temperatures. According to the images, the best coverage was achieved at 100 °C which support previous report [6]. While at 175 °C the coverage was not as good as at 100 °C. At 230 °C, the $MAPbI_3$ decomposes to PbI_2 as can be seen by the XRD analysis shown in Fig. 5.5d (purple line) which was performed to the sample after annealing for 15 min at 230 °C. The mixture (Fig. 5.6b, e, h) shows good coverage for all annealing temperatures. The presence of the two perovskite crystals, $MAPbI_3$ and the $FAPbI_3$, are observed in the inset of Fig. 5.6e, corresponding to 175 °C annealing temperature which further support the XRD spectra of the mixture as shown in Fig. 5.5d. For the $FAPbI_3$ (Fig. 5.6c, f, i), the crystals are hardly recognizable at 100 °C while at 175 and 230 °C the $FAPbI_3$ crystals are observed clearly, supporting the required high annealing temperature of the $FAPbI_3$ perovskite.

Further investigation of the dependence of $FAPbI_3$ and $MAPbI_3$ films at various temperatures can be seen in Fig. 5.6j. Seven different temperatures (100, 150, 180, 200, 230, 260, and 290 °C) were tested. Early reports showed that in the two-step deposition method, an annealing temperature of ∼170 °C [11] is used for the $FAPbI_3$ perovskite, while for $MAPbI_3$, the annealing temperature is 70–100 °C.

As indicated in Fig. 5.6j, the MAPbI$_3$ perovskite has a dark color until 200 °C, while at ∼230 °C the MAPbI$_3$ started to decompose. On the other hand, the FAPbI$_3$ perovskite has a dark color till 260 °C while at 290 °C it decomposes. According to these results, an annealing temperature of ∼175 °C is suitable for a mixture of FAPbI$_3$ with MAPbI$_3$. This observation is in good agreement with the XHR-SEM images. Consequently, it is reasonable that the side reactions (such as sublimation, evaporation and decomposition of MAPbI$_3$ and FAPbI$_3$) could take place in the mixture samples. The direct observation of these reactions is that despite the fact that the original molar ratio of MAPbI$_3$:FAPbI$_3$ in the dipping solution was 1:1, it can be that the actual molar ratio between MAPbI$_3$ and FAPbI$_3$ after the annealing process is not 1:1. According to the stability test we performed, it is reasonable to estimate that the portion of FAPbI$_3$ is a bit larger than that of MAPbI$_3$. MAPbI$_3$, which owns the smaller cation, is more likely to be affected by the high temperature. In other words, the side reaction could be expected to diminish the MAPbI$_3$ content more dramatically than the FAPbI$_3$ content.

Figures 5.7a, b present the photovoltaic parameters of the complete hole-conductor-free MAPbI$_3$, FAPbI$_3$ and the mixture (MAPbI$_3$:FAPbI$_3$) solar cells were measured at eight different temperatures (22, 35, 45, 55, 65, 75, 85, 95 °C).

The general trend that is reflected from this analysis is similar for all the cells. The J_{sc} of all samples is increased as temperature increases. Increasing the temperature of a semiconductor results with a decreasing of E_g which could influence the J_{sc}. It can be seen that the J_{sc} increases by more than 30 % in the case of FAPbI$_3$. This is expected due to the better stability of FAI at high temperatures. The MAPbI$_3$ has an initial increase of 20 % in J_{sc} while the FAPbI$_3$ stays stable for the rest of the temperature range. An interesting point is that the cells which used a mixture of MAPbI$_3$ and FAPbI$_3$ showed the most severe degradation of the PV parameters with the increase in temperature. A possible explanation for this phenomenon is that the increase in temperature harmed the crystals from the inside.

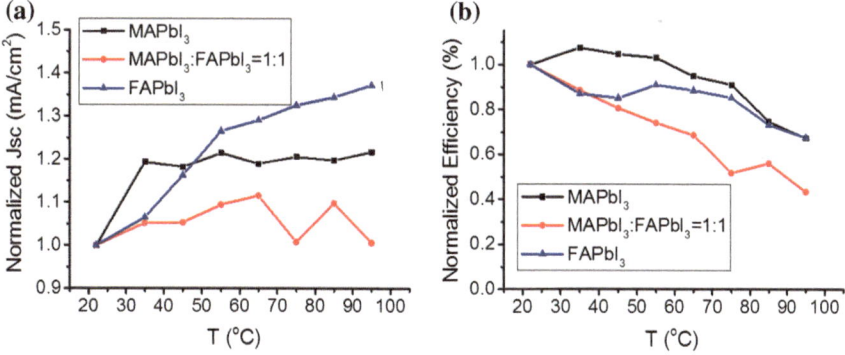

Fig. 5.7 **a** The normalized short-circuit current density (J_{sc}) of the three different photovoltaic cells at various temperatures. **b** The normalized efficiency of the photovoltaic cells at various temperatures. Taken with permission from Ref. [7]

The active layer included a combination of MA$^+$ and FA$^+$ in the crystal voids. The two cations are affected differently by the temperature increase, and this difference could lead to the formation of strain inside the crystalline structure. Such a strain could be harmful when measuring the photovoltaic performance of the cell at varying temperatures. In contrast to the temperature dependence on J_{sc}, the efficiency decreased as the temperature was increased (Fig. 5.7b). This behavior is due to the decrease in the FF and the V_{oc}. When increasing the temperature, the recombination increases, resulting in a decrease of the V_{oc}. Additional contribution to the decrease of the V_{oc} can be related to the increase of the intrinsic carrier when the temperature increases. The intrinsic carrier concentration depends on the band gap, lowering the band gap giving higher intrinsic concentration which can result of lower V_{oc}.

Also in this case, the mixture showed the lowest results regarding temperature dependence. The increase in J_{sc} does not compensate for the decrease in the FF and the V_{oc}, and as a result, the efficiency decreases.

Intensity modulated photocurrent/photovoltage spectroscopies were used to measure the transport and recombination times (t and r, respectively) of the solar cells studied. The electron diffusion length L_d in the perovskite solar cell can be determined using the expression [12–14] $L_d = (\tau_r D)^{1/2}$ where D is the diffusion coefficient and τ_r is the recombination lifetime. The electron diffusion length represents the average distance which electron travels before it recombines by holes in the perovskite or at the back contact. Figure 5.8a shows the diffusion length

Fig. 5.8 a Diffusion length (L_d) as a function of light intensity for the cells studied. **b** Charge collection efficiency and recombination lifetime (τ_r) as a function of light intensity. The *dashed lines* correspond to the recombination lifetimê. Taken with permission from Ref. [7]

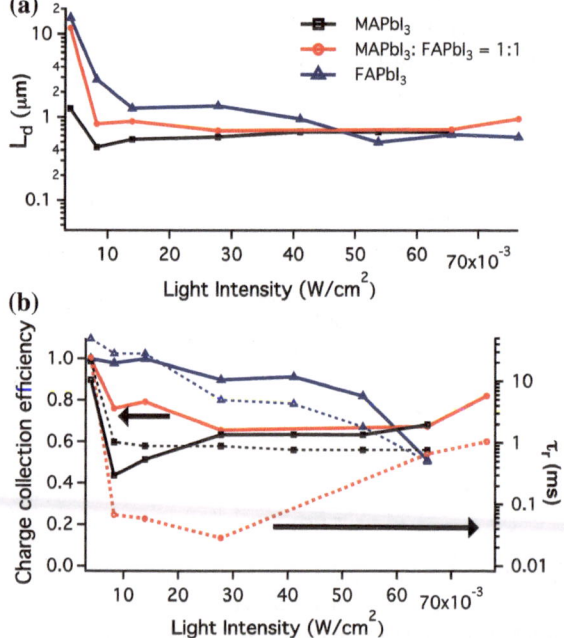

calculated using IMVS and IMPS as a function of the light intensity. At low light intensity, all cells have higher diffusion length, although the $FAPbI_3$ and the mixture have longer initial diffusion length. When increasing the light intensity, all cells decreased to the same diffusion length of around 0.8–1 μm (this result is in good agreement with literature). There is weak light intensity dependence of the diffusion length. The diffusion length varies by less than a factor of 1 over more than seven decades of light intensity, similar to the dependence of diffusion length by the light intensity in dye sensitized solar cells [15].

The charge collection efficiency and the recombination lifetime are presented in Fig. 5.8b. The charge collection efficiency can be calculated using [16] $\eta_{cc} = 1 - \tau_t/\tau_r$. The $FAPbI_3$ cell starts with high η_{cc} at low light intensity and drops to 0.5 at higher intensities. On the other hand, the $MAPbI_3$ and the mixture have low η_{cc} at low intensities, which increases to 0.7–0.8 at higher light intensities. These results explain well the PV performance of these cells at one sun illumination. Moreover, the recombination lifetime (presented as dashed lines in Fig. 5.8b) decreases when increasing the light intensity as expected. However, for the $MAPbI_3$ cell, the recombination lifetime remains constant for most of the intensity ranges which supports the high efficiency observed for the $MAPbI_3$-based cells.

One of the important observations till recently are that mixed halides or cations in the perovskite structure are beneficial for the PV performance and for the cells stability. Meaning that the halide or the cation concentration will not be more than 20 % in the mixture relative to the other halide or cation. This has a meaningful in stabilizing the correct OMHP phase.

References

1. Aharon S, Cohen B-E, Etgar L (2014) Hybrid lead halide iodide and lead halide bromide in efficient hole conductor free perovskite solar cell. J Phys Chem C 118:17160–17165
2. Abrusci A, Stranks SD, Docampo P, Yip H-L, Jen AKY, Snaith HJ (2013) High-performance perovskite-polymer hybrid solar cells via electronic coupling with fullerene monolayers. Nano Lett 13:3124–3128
3. Kitazawa N (1997) Excitons in two-dimensional layered perovskite compounds: $(C_6H_5C_2H_4NH_3)_2Pb(Br, I)_4$ and $(C_6H_5C_2H_4NH_3)_2Pb(Cl, Br)_4$. Mater Sci Eng B 49:233–238
4. Ishihara T, Takahashi J, Goto T (1990) Optical properties due to electronic transitions in two-dimensional semiconductors $(C_nH_{2n} + 1NH_3)PbI_4$. Phys Rev B 42(17):11099
5. Ishihara T (1994) Optical properties of PbI-based perovskite structures. J Lumin 60–61:269–274
6. Aharon S, Gamliel S, Cohen B, Etgar L (2014) Depletion region effect of highly efficient hole conductor free $CH_3NH_3PbI_3$ perovskite solar cells. Phys Chem Chem Phys 16:10512–10518
7. Aharon S, Dymshits A, Rotem A, Etgar L (2015) Temperature dependence of hole conductor free formamidinium lead iodide perovskite based solar cells. J Mater Chem A 3:9171–9178
8. Supasai T, Rujisamphan N, Ullrich K, Chemseddine A, Dittrich T (2013) Formation of a passivating $CH_3NH_3PbI_3/PbI_2$ interface during moderate heating of $CH_3NH_3PbI_3$ layers. Appl Phys Lett 103:183906
9. Barnea-Nehoshtan L, Kirmayer S, Edri E, Hodes G, Cahen D (2014) Surface photovoltage spectroscopy study of organo-lead perovskite solar cells. J Phys Chem Lett 5:2408–2413

10. Kronik L, Shapira Y (2001) Surface photovoltage spectroscopy of semiconductor structures: at the crossroads of physics, chemistry and electrical engineering. Surf Interface Anal 31:954–965
11. Epron GE, Stranks SD, Manelaou C, Johnston MB, Herz LM, Snaith HJ (2014) Formamidinium lead trihalide: a broadly tunable perovskite for efficient planar heterojunction solar cells. Energy Environ Sci 7:982–988
12. Frank AJ, Kopidakis N, van de Lagemaat J (2004) Electrons in nanostructured TiO_2 solar cells: transport, recombination and photovoltaic properties. Coord Chem Rev 248:1165–1179
13. Zhao Y, Nardes AM, Zhu K (2014) Solid-state mesostructured perovskite $CH_3NH_3PBI_3$ solar cells: charge transport, recombination, and diffusion length. J Phys Chem Lett 5:490–494
14. Zhao Y, Zhu K (2013) charge transport and recombination in perovskite (CH3NH3)PbI3 sensitized TiO_2 solar cells. J Phys Chem Lett 4:2880–2884
15. Peter LM, Wijayantha KGU (2000) Electron transport and back reaction in dye sensitized nanocrystalline photovoltaic cells. Electrochim Acta 45:4543–4551
16. Zhu K, Jang S-R, Frank AJ (2011) Impact of high charge-collection efficiencies and dark energy-loss processes on transport, recombination, and photovoltaic properties of dye-sensitized solar cells. J Phys Chem Lett 2:1070–1076
17. Mosconi E, Amat A, Nazeeruddin MK, Graïzel M, Angelis FD (2013) First-principles modeling of mixed halide organometal perovskites for photovoltaic applications. J Phys Chem C 117:13902–13913

Chapter 6
High Voltage in Hole Conductor Free Organo Metal Halide Perovskite Solar Cells

One of the attractive properties of organo metal halide perovskite is its ability to gain high open-circuit voltage (V_{oc}) with a high ratio of qV_{oc}/E_g [1]. Several works demonstrate high V_{oc} in perovskite-based solar cells which have hole transport material layer (HTM). HTMs such us: poly[N-9-hepta-decanyl-2,7-carbazole-alt-3,6-bis-(thiophen-5-yl)-2,5-dioctyl-2,5-dihydropyrrolo[3,4-]pyrrole-1,4-dione] (PCBTDPP), N,N'-dialkylperylenediimide (PDI), 4,4'-bis(N-carbazolyl)-1,1'-biphenyl (CBP), and tri-arylamine (TAA) polymer derivatives containing fluorene and indenofluorene were demonstrated to achieve high V_{oc} between 1.15 and 1.5 V [2–5].

All the reports mentioned relating to high voltage based on perovskite use hole transport material to tune and to gain high V_{oc}. However, is it possible to get high V_{oc} without hole transport material?

Based on recent reports [4, 5], the voltage in the perovskite solar cells is not determined simply by the difference between the TiO_2 fermi level and the fermi level of the HTM. Therefore, it could be that high V_{oc} can be achieved in perovskite-based solar cells even without HTM.

The work below showed the ability to gain V_{oc} of up to 1.35 V without the use of HTM. The solar cell structure used in this work has the same configuration as described earlier for HTM free cell. When two perovskite structures were studied: $CH_3NH_3PbI_3$ or $CH_3NH_3PbBr_3$ (CH_3NH_3 = MA) deposited by the two-step deposition. Moreover, two kinds of mesoporous metal oxides were studied TiO_2 and Al_2O_3 NPs.

Table 6.1 present the photovoltaic parameters achieved for the high voltage cells. Four different combinations were studied—nanocrystalline TiO_2 with MAPbI$_3$ and MAPbBr$_3$, and nanocrystalline Al_2O_3 with both perovskites. The open-circuit voltage (V_{oc}) for the cells with the MAPbBr$_3$ perovskite deliver higher voltages compared with the cells with the MAPbI$_3$ perovskite related to the same metal oxide. Moreover, the cells with Al_2O_3 achieve higher V_{oc} compared to the TiO_2-based cells. The highest V_{oc} observed for the Al_2O_3/MAPbBr$_3$ configuration achieved 1.35 V without a hole conductor. This is the highest reported open-circuit

© The Author(s) 2016
L. Etgar, *Hole Conductor Free Perovskite-based Solar Cells*,
SpringerBriefs in Applied Sciences and Technology,
DOI 10.1007/978-3-319-32991-8_6

Table 6.1 Photovoltaic parameters of the hole conductor free solar cells studied

	η (%)	Fill factor	J_{sc} (mA/cm^2)	V_{oc} (V)
TiO$_2$/MAPbI$_3$	7.5	0.61	14.1	0.86
TiO$_2$/MAPbBr$_3$	1.88	0.49	4.37	0.87
Al$_2$O$_3$/MAPbI$_3$	4.13	0.5	7.46	1.0
Al$_2$O$_3$/MAPbBr$_3$	2.02	0.55	2.7	1.35

voltage for perovksite cells without a hole conductor and is comparable to cells using hole transport material. It is important to note that the average V_{oc} for the Al$_2$O$_3$/MAPbBr$_3$ configuration is 1.24 ± 0.08 V with 4 cells having V_{oc} of more than 1.3 V and 4 cells with V_{oc} higher than 1.21 V.

High power conversion efficiency (PCE) with high voltage was observed for the Al$_2$O$_3$/MAPbI$_3$ configuration achieving PCE of 4.1 % with V_{oc} of 1 V.

Figure 6.1a, b show the energy level diagram for the four different cases described in this paper. Figure 6.1a presents the MAPbBr$_3$ and MAPbI$_3$ deposited on Al$_2$O$_3$ which function as a scaffold; electron injection from the perovskite to the Al$_2$O$_3$ is not possible in this configuration. Figure 6.1b presents the MAPbBr$_3$ and MAPbI$_3$ with TiO$_2$, where electron injection from the perovksite to the TiO$_2$ is favorable. Surface photovoltage (SPV) was used to measure the work function of the Al$_2$O$_3$, TiO$_2$, MAPbI$_3$ and MAPbBr$_3$; the calculated work functions are shown as red lines in Fig. 6.1. The work function positions (which are the Fermi level positions) correspond well to the p-type behavior of the MAPbBr$_3$ and MAPbI$_3$ perovskites, and the n-type behavior of the TiO$_2$ as discussed below. Figure 6.1c, d show the corresponding high-resolution scanning electron microscopy (HR-SEM) cross sections of the Al$_2$O$_3$/MAPbI$_3$ and Al$_2$O$_3$/MAPbBr$_3$ HTM free cells, the perovskite over layer can be observed clearly.

Surface photovoltage (SPV) spectroscopy and incident-modulated photovoltage spectroscopy (IMVS) were performed to gain more information about the reason for the high open-circuit voltage when no HTM is used. Previous studies already demonstrate high V_{oc} in perovskite cells contain HTM. It was suggested that the V_{oc} is not merely the difference between the hole Fermi level of the hole conductor and the electron Fermi level of the nanocrystalline TiO$_2$ [6]. Moreover, it was reported that charges could be accumulated in the perovskite due to its high capacitance, which allows the control of the quasi Fermi level during illumination [7].

The SPV spectra of the MAPbI$_3$ and MAPbBr$_3$ are shown in Fig. 6.2a, with the estimated band gaps shown as vertical lines. Several observations result from the SPV spectra (Fig. 6.2a). First, the spectra provide information about the band gaps of the materials, equivalent to the information observed from the absorption spectra. Second, the sign of the SPV signal indicates the samples type. The surface work function is changed on illumination. It decreases for the n-type semiconductor—TiO$_2$ in this case—and increases for the p-type semiconductor, the MAPbI$_3$ and MAPbBr$_3$ in this case. A third important observation is related to the unique property of the SPV method, its immunity to reflection or scattering losses only

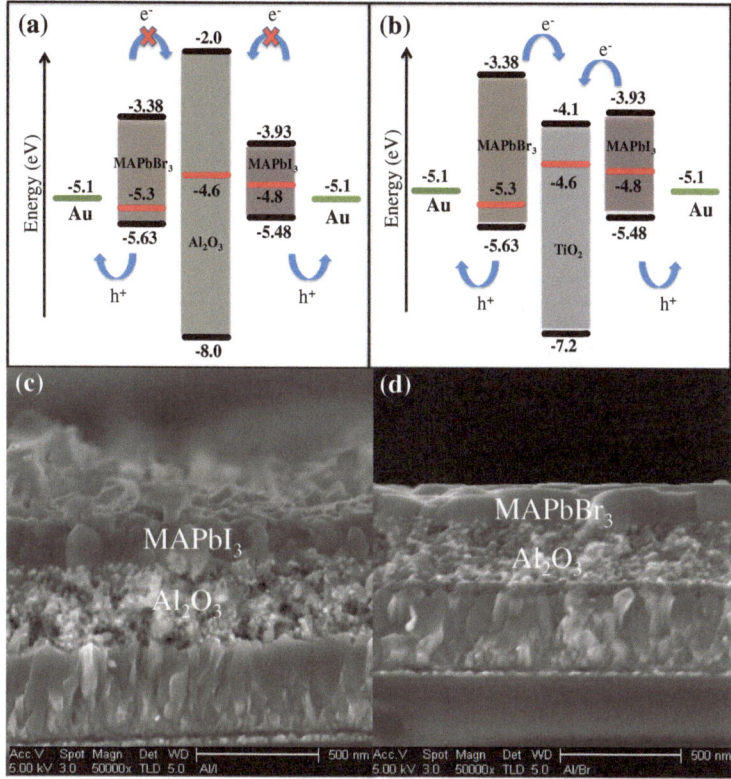

Fig. 6.1 a, b Energy level diagram of the different cells. Fermi levels measured under dark are presented in *red* in the figure. The position of the conduction and valence bands are according to Ref. [11]. **c** HR-SEM cross section of Al₂O₃/MAPbI₃ HTM free cell. **d** HR-SEM cross section of Al₂O₃/MAPbBr₃ HTM free cell. Taken with permission from Ref. [1]

photons that are absorbed in the sample contribute to the SPV signal. In this case, the signal onset starts at photon energies very close to the band gap of the perovskite samples (the MAPbI₃ and the MAPbBr₃), and as a result, it can be concluded that the perovskite samples have fewer sub-band gap states. The relation qV_{oc}/E_g is the ratio of the maximum voltage developed by the solar cell (V_{oc}) to the voltage related to the band gap of the absorber (E_g/q). For the Al₂O₃/MAPbBr₃ cell the qV_{oc}/E_g is 0.61, compared to recent reports of high voltage perovskite cells with HTM shown as qV_{oc}/E_g of 0.55 and 0.73 [3]. Our results demonstrate comparable values with respect to the values obtained with HTM.

Calculating the qV_{oc}/E_g relation, it is found that the Al₂O₃/MAPbBr₃-based cells have slightly more thermal losses than the Al₂O₃/MAPbI₃ cells, moreover high thermal losses were observed in the TiO₂-based cell compared to the Al₂O₃-based cells.

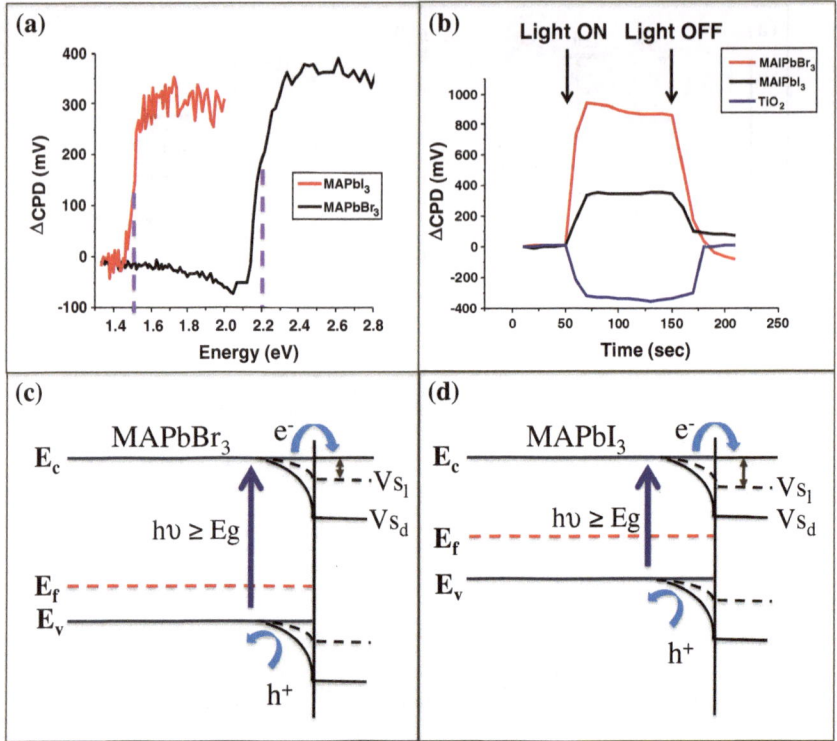

Fig. 6.2 **a** SPV spectra of the MAPbI$_3$ and MAPbBr$_3$ films with estimated band gaps, 1.55 and 2.2 eV, respectively. **b** Contact potential difference (CPD) change with white light switched on and off for the samples studied. **c** The effect of band-to-band transitions on the surface photovoltage responses of MAPbBr$_3$. **d** MAPbI$_3$. V_{s_d}—surface potential in the dark, V_{s_l}—surface potential in light. Taken with permission from Ref. [1]

Figure 6.2b presents the contact potential difference (CPD) change when a light is switched on and off. The SPV onset t_{on} and t_{off} are below the resolution limit of the measurements system. Our observations from these measurements are related to the change in the CPD—for the TiO$_2$ sample, a negative change in the CPD was observed, while for the perovskites samples, a positive change in the CPD was observed, corresponding to their electronic behavior. The ΔCPD for the MAPbI$_3$ is 350 mV and the ΔCPD for the MAPbBr$_3$ is 850 mV.

The V_S presented in Fig. 6.2c, d is the surface potential barrier (where V_{s_d} is the surface potential in the dark and V_{s_l} is the surface potential in light), which is measured in the SPV experiment. The difference between the surface potential in the light (V_{s_l}) and in the dark (V_{s_d}) is defined as the SPV signal. In super band gap illumination, photons with energy equal or larger than the band gap hit the material and generate electron-hole pairs, which are collected by the surface barrier, consequently the surface potential is reduced. The trap to band transition is neglected in super band gap illumination while the band-to-band absorption is the dominant one.

Figure 6.2c, d show the effect of band-to-band transition on the SPV response of MAPbBr$_3$ and MAPbI$_3$, respectively. Under illumination, there is redistribution of surface charges, which decrease the bend bending, and as a result the SPV response is generated. According to the ΔCPD shown in Fig. 6.2b, the surface potential for the MAPbBr$_3$ is smaller than the surface potential for the MAPbI$_3$ as indicated by the bidirectional arrow in Fig. 6.2c, d, respectively. (This is also seen by the bend bending reduction, since the difference between V_{s_l} and V_{s_d} is larger in the case of MAPbBr$_3$). The reduction of the surface potential observed from the SPV measurements for the HTM free MAPbBr$_3$ cells, could be a possible contribution to the higher V_{oc} achieved for these cells.

Further contribution for the difference in the open-circuit voltage is presented in Fig. 6.3. The recombination lifetime (τ_r) as a function of the voltage was calculated by IMVS [8–10]. All cells showed the same dependence of electron recombination lifetime (which are the minority carriers) by the voltage, the decrease of the recombination lifetime with increasing the voltage. This behavior can be attributed to the increased recombination with the higher electron density. However, the τ_r values are different for the various cells; in particular, the lower τ_r values were observed for the TiO$_2$/MAPbI$_3$ cell, which also had the lower V_{oc}. The highest τ_r values were observed for the Al$_2$O$_3$/MAPbBr$_3$ cell, which result with less recombination (since the lifetime for recombination is longer) corresponding with the highest voltage observed. Longer recombination lifetime (τ_r) will decrease the recombination, which could result in higher V_{oc}. In addition, longer recombination lifetimes were observed for the cells with the Al$_2$O$_3$ as a scaffold, compared to cells with mesoporous TiO$_2$, which could contribute to the higher voltages observed in the case of Al$_2$O$_3$-based cells. This is consistent with the qV_{oc}/E_g relation as

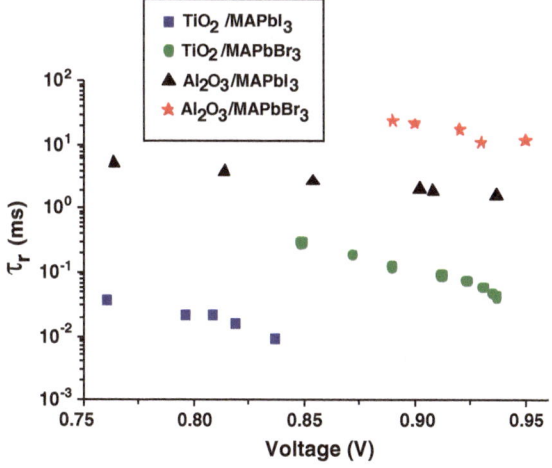

Fig. 6.3 Recombination lifetime (τ_r) as a function of the open-circuit voltage for the cells studied, measured by IMVS. Taken with permission from Ref. [1]

discussed above, in which more thermal losses were observed for cells based on TiO_2 as the metal oxide compared to cells based on Al_2O_3.

Summarizing this chapter, V_{oc} of 1.35 V was observed for perovskite-based cells without HTM. It can be concluded that the V_{oc} is not solely influenced by the energy levels of the HTM and the main origin for the V_{oc} in perovskite-based solar cells is attributed to the band gap of the material and to the interface with the metal oxide.

References

1. Dymshits A, Rotem A, Etgar L (2014) High voltage in hole conductor free organo metal halide perovskite solar cells. J Mater Chem A 2:20776–20781
2. Cai B, Xing Y, Yang Z, Zhang WH, Qiu J (2013) High performance hybrid solar cells sensitized by organolead halide perovskites. Energy Environ Sci 6:1480
3. Edri E, Kirmayer S, Cahen D, Hodes G (2013) High open-circuit voltage solar cells based on organic-inorganic lead bromide perovskite. J Phys. Chem Lett 4:897–902
4. Edri E, Kirmayer S, Kulbak M, Hodes G, Cahen D (2014) Chloride inclusion and hole transport material doping to improve methyl ammonium lead bromide perovskite-based high open-circuit voltage solar cells. J Phys Chem Lett 5:429–433
5. Ryu S, Noh JH, Jeon NJ,Kim YC, Yang WS, Seo J, Seok SI (2014) Voltage output of efficient perovskite solar cells with high open-circuit voltage and fill factor. Energy Environ Sci. doi:10.1039/c4ee00762j
6. Chiang YF, Jeng JY, Lee MH, Peng SR, Chen P, Guo TF, Wen TC, Hsu YJ, Hsu CM (2014) High voltage and efficient bilayer heterojunction solar cells based on an organic–inorganic hybrid perovskite absorber with a low-cost flexible substrate. Phys Chem Chem Phys 16:6033
7. Kim H-S, Mora-Sero I, Gonzalez-Pedro V, Fabregat-Santiago F, Juarez-Perez EJ, Park N-G, Bisquert J (2013) Mechanism of carrier accumulation in perovskite thin-absorber solar cells. Nat commun. doi:10.1038/ncomms3242
8. Zhu K, Jang S-R, Frank AJ (2011) Impact of high charge-collection efficiencies and dark energy-loss processes on transport, recombination, and photovoltaic properties of dye-sensitized solar cells. J Phys Chem Lett 2:1070–1076
9. Zhao Y, Zhu K (2013) Charge transport and recombination in perovskite (CH_3NH_3) PbI_3 sensitized TiO_2 solar cells. J Phys Chem Lett 4:2880–2884
10. Zhao Y, Nardes AM, Zhu K (2014) Solid-state mesostructured perovskite $CH_3NH_3PbI_3$ solar cells: charge transport, recombination, and diffusion length. J Phys Chem Lett 5:490–494
11. Hao F, Stoumpos CC, Chang RPH, Kanatzidis MG (2014) Anomalous band gap behavior in mixed Sn and Pb perovskites enables broadening of absorption spectrum in solar cells. J Am Chem Soc 136:8094–8099

Chapter 7
Self-assembly of Perovskite for Fabrication of Semi-transparent Perovskite Solar Cells

The last topic discussed in this book is related to the possibility and to the idea of forming semi-transparent perovskite solar cells using self-assembled process which enables the control over the solar cell transparency [1].

There are not many reports on semi-transparent perovskite-based solar cells. Eperon et al. [2, 3] used a strategy which relied on the dewetting of the perovskite film to create 'perovskite islands', thus achieving high transmittance but clearly decreasing the overall PCE due to the voids within the active layer. Moreover, it seems that with this method it is difficult to precisely control the transparency of the cell. In recent reports [4, 5], a thin perovskite layer was deposited by evaporation technique. However, evaporation-based processes are very costly, require high capital investments, and are very complicated for upscaling, which is required for industrial applications. Semi-transparent top electrode made of silver nanowires was introduced into perovskite-based solar cells [6]. The transparency in this case was controlled by the top electrode transparency and not by the perovskite film. In addition, the silver nanowires can be used as an alternative top electrode made by solution-processed technique for semi-transparent solar cells.

Here we report on a unique, simple wet deposition method for the fabrication of semi-transparent perovskite-based solar cells. This deposition method is fundamentally different from previously reported deposition methods of $CH_3NH_3PbI_3$ ($MAPbI_3$) perovskite. The film formation in this method is enabled by the mesh-assisted assembly of the perovskite solution through wetting along the wall of a conventional screen printing mesh. Meaning, here the perovskite is deposited along a *controlled pattern* through *solution-process* and *in ambient conditions*.

The structure of the perovskite-based solar cells is as follows: FTO glass/TiO_2 compact layer/mesoporousTiO_2/grid of Perovskite/spiro-OMeTAD/thin film of gold (Fig. 7.1a). In the case of the HTM-free semi-transparent perovskite solar cells, the structure was identical excluding the spiro-OMeTAD layer (Fig. 7.1b). Due to the controlled voids in the perovskite grid, which enables the transparency of the solar cell, a direct contact is formed between the HTM and the mesoporous TiO_2 film (or between the gold back contact and the mesoporous TiO_2 film, in the

© The Author(s) 2016
L. Etgar, *Hole Conductor Free Perovskite-based Solar Cells*,
SpringerBriefs in Applied Sciences and Technology,
DOI 10.1007/978-3-319-32991-8_7

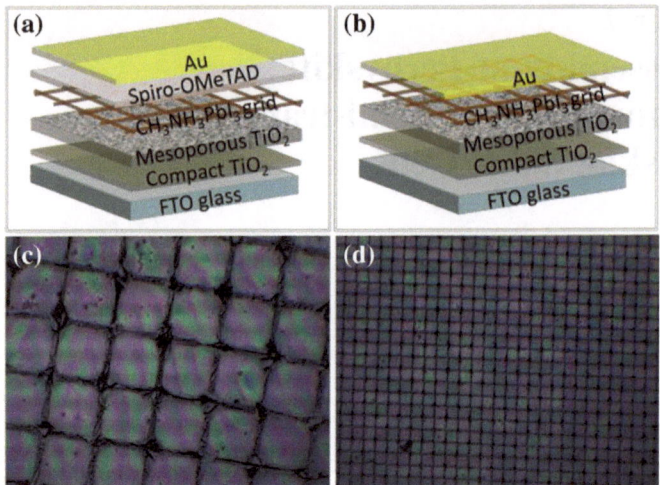

Fig. 7.1 a The structure of the semi-transparent cell: FTO glass/TiO$_2$ compact layer/mesoporous TiO$_2$/grid of CH$_3$NH$_3$PbI$_3$/Spiro-OMeTAD/gold. **b** The structure of the HTM-free semi-transparent cell: FTO glass/TiO$_2$ compact layer/mesoporous TiO$_2$/grid of CH$_3$NH$_3$PbI$_3$/gold. **c** Optical microscope image of the wide grid: opening of ∼200 μm. **d** Optical microscope image of the dense grid: opening of ∼60 μm. Taken with Permission form Ref. [1]

case of HTM-free solar cells). Interestingly, despite those voids, the semi-transparent devices with and without HTM showed power conversion efficiency of 4.98 and 2.55 %, respectively, for 19 % transparency. The perovskite grid was obtained through wetting of the perovskite precursor solution along the walls of a screen printing mesh which was placed on the mesoporous TiO$_2$ layer. As shown below, the transparency of the cell can be simply controlled by two parameters, (i) mesh opening, and (ii) the precursor solution concentration.

Following the deposition of the compact and mesoporous TiO$_2$ layers, a solution of PbI$_2$ and MAI in DMF was placed on top of the screen printing mesh. Due to the presence of the wetting agent, the solution immediately wets the surface and fills the gap between the screen printing mesh and the mesoporous TiO$_2$ layer. During solvent evaporation, the PbI$_2$ and MAI molecules aligned along the mesh wires due to capillary forces. The screen printing mesh was removed after the DMF evaporation was completed, and a grid pattern was obtained, in which the walls contain crystalline perovskite formed in situ from PbI$_2$ and MAI. An additional annealing step is necessary to complete the crystallization.

Figure 7.1c, d present optical microscope images of two different CH$_3$NH$_3$PbI$_3$ perovskite grids fabricated on top of the mesoporous TiO$_2$ surface using meshes with different openings: wide opening (Fig. 7.1c, grid opening of ∼200 μm), and narrow opening (Fig. 7.1d, grid opening of ∼60 μm). In general, it was found that the grid opening size can be easily controlled by the mesh opening size, thus enabling control of the transparency of the perovskite film (which is actually empty

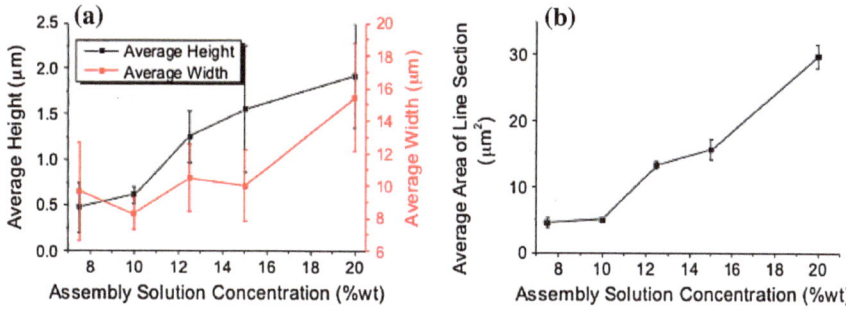

Fig. 7.2 The grid dimensions versus the assembly solution concentration (measured by profiler). **a** The average height and width (μm, Full Width at Half Maximum, FWHM) of the grid lines resulted from use of different concentrations of the assembly solutions (wt%). **b** The average cross section area of a grid line (calculated by multiplying the average width line with the average height of the same grid line) versus assembly solution concentration. Taken with Permission form Ref. [1]

squares surrounded by perovskite crystals). The power conversion efficiency (PCE) is affected by the grid opening, i.e., when the grid opening is large, the PCE is decreased since the active area is smaller and there are more contact points between the hole transport material (HTM) and the mesoporous TiO_2. Moreover, no perovskite crystals are present at the open areas in the perovskite grid as shown in Fig. 7.1c, d. The absence of perovskite crystals in the voids supports the efficient control over the final assembly of the perovskite attained using this deposition method.

As seen in Fig. 7.2a, the average height (relative to the TiO_2 layer) of the grid line increases while increasing the solution concentration. However, the width has the same values when using concentrations between 7.5 and 15 wt%, and starts to increase only above this concentration. There is a minimal line width which is a result of the capillary forces between the perovskite molecules and the mesoporous TiO_2 layer, meaning, even at low concentrations this width is expected to be covered by molecules. Figure 7.2b shows the average area of the grid line cross section, (calculated by multiplying the average width of a specific grid line with the average height of the same grid line). As expected, the area of the grid line cross section is increased while increasing the concentration of the components of the perovskite precursor solution.

The current-voltage curves of various assembly solution concentrations using dense grids with opening of ~60 μm are presented in Fig. 7.3d. The PCE of the cells decreases with the increase in transparency (Fig. 7.3a), which is controlled by the precursor solution concentration (Fig. 7.3b). This is expected since the grid lines which are formed by the low concentration solution have smaller average widths and heights than those of the grid lines formed from a more concentrated assembly solution. Moreover, Fig. 7.3a, b show the reproducibility of the different cells using various concentrations. The efficiencies and the transparencies are more widely distributed when increasing the solution concentration.

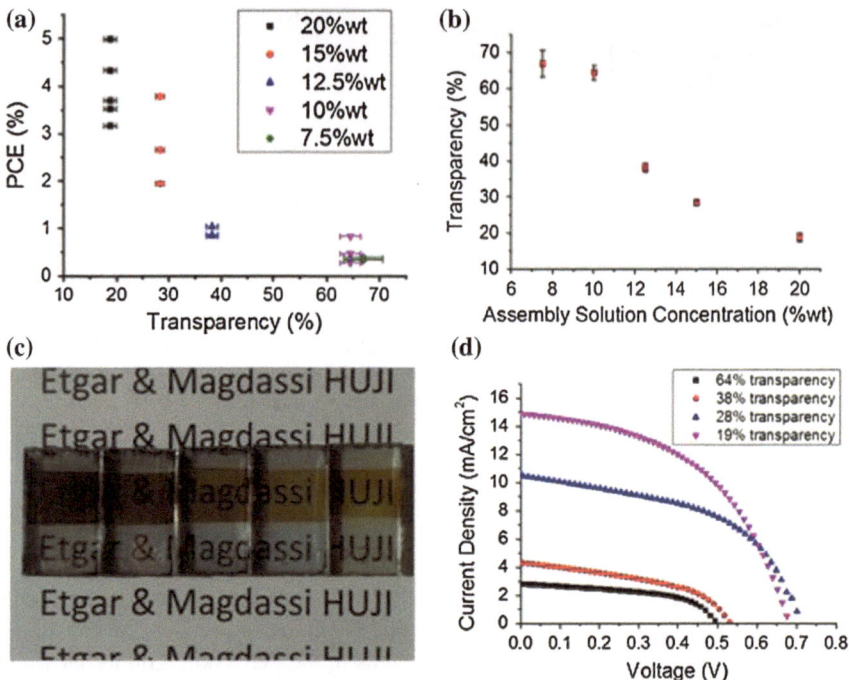

Fig. 7.3 a The PCE of the semi-transparent cells versus their average transparency calculated in the range of 400–800 nm wavelengths (the different colors represent the different solution concentrations). **b** The average transparency of the perovskite solar cells versus the concentration (wt%) of the assembly solution of the precursors. **c** Images of the semi-transparent cells with different average transparencies from left to right,19, 28, 38, 64, and 67 %. **d** The current-voltage curves of the semi-transparent cells with different average transparencies. The PCEs of the cells are 0.83, 1.04, 3.79, and 4.98 % for the cells of 64, 38, 28 and 19 % transparency, respectively. Taken with Permission form Ref. [1]

Figure 7.3c shows the images of the cells with different transparencies (which were controlled by changing the assembly concentration solution from 7.5 to 20 wt %, from right to left).

The transparency control by the perovskite grid formation has an important advantage, which is the possibility to calculate the exact coverage of the perovskite grid in the active area. Using this unique mesh-assisted assembly deposition method, it is possible to accurately design the required coverage and transparency of the solar cell. The coverage of the cell is the fraction of area that is covered by the perovskite grid lines, without the grid voids (in which there is no perovskite). It was observed that the coverage of the perovskite solar cells prepared using assembly solution concentrations of 7.5, 10, 12.5, and 15 wt% are similar ($\sim 35 \pm 3$ % of the area is covered by the perovskite grid lines), while only the coverage of cells that were prepared using assembly solution concentration of 20 wt % the coverage was different (~ 50 %). The increase in the coverage for the 20 wt

% solution concentration can be expected already from the profile measurements (Fig. 7.3a) in which the width of the grid lines was about the same average values in the case of the lower concentrations (till 15 wt%). However, despite the fact that the coverage of the cells in the case of the lower assembly solution concentrations was similar, the cells PV performance was different, as shown in Fig. 7.3d. The reason for the difference in PV performance might be due to the fact that the coverage calculations do not take into account the thickness of the perovskite lines, which dramatically affects the PV performance.

References

1. Aharon S, Layani M, Cohen BE, Shukrun E, Magdassi S, Etgar L (2015) Self-assembly of perovskite for fabrication of semi-transparent perovskite solar cells. Adv Mater Interf (Accepted)
2. Eperon GE, Burlakov VM, Goriely A, Snaith HJ (2014) Neutral color semitransparent microstructured provskite solar cells. ACS Nano 8(1):591–598
3. Eperon GE, Bryant D, Troughton J, Stranks SD, Johnston MB, Watson T, Worsley DA, Snaith HJ (2015) Efficient, semitransparent neutral-colored solar cells based on microstructured formamidinium lead trihalide perovskite. J Phys Chem Lett 6:129–138
4. Roldan C, Malinkiewicz O, Betancur R, Longo G, Momblona C, Jaramillo F, Camacho L, Bolink HJ (2014) High efficiency single-junction semitransparent perovskite solar cells. Energy Environ Sci. doi:10.1039/C4EE01389A
5. Ono LK, Wang S, Kato Y, Rega SR, Qi Y (2014) Fabrication of semi-transparent perovskite films with centimeter-scale superior uniformity by the hybrid deposition method. Energy Environ Sci
6. Guo F, Azimi H, Hou Y, Przybilla T, Hu M, Bronnbauer C, Langner S, Spiecker E, Forbericha K, Brabec CJ (2015) High-performance, semitransparent perovskite solar cells with solution processed silver nanowires as top electrods. Nanoscale 7:1642–1649

Chapter 8
Summary

Organo-Metal halide perovskite (OMHP) are a new class of materials that have been used in optoelectronic applications. This brief book discusses their exciting properties in particular their ability to conduct electron and holes efficiently in the solar cell.

The OMHP is made by solution processed which might be the major advantage for being so attractive and promising research and applicative field.

Along with this book, several critical subjects related to perovskite solar cells were discussed. At the beginning, the parameters that influence the OMHP deposition were investigated as described in Chap. 4. Following the understanding of these parameters, we were able to investigate the working mechanism of this hole conductor (HTM) free solar cell structure which reveals a depletion region at the TiO_2/OMHP interface. Increasing the depletion region will suppress recombination resulting with enhanced power conversion efficiency.

In order to further improve the performance of the OMHP HTM free solar cell, an anti-solvent treatment was applied onto the OMHP surface. The anti-solvent treatment changes the film morphology and improves the film coverage, which is beneficial for the performance of these HTM free cells. The conductive atomic force microscopy and surface photovoltage techniques show that the electronic properties of the perovskite film also change due to the anti-solvent treatment: the perovskite film became slightly more intrinsic, further contributing to the enhanced performance. During the toluene treatment, halide and methylammonium ions are removed from the surface which creates a net positive charge on the Pb atoms, resulting with more conductive surface of the perovskite, which is beneficial for the HTM free solar cell structure resulting with efficiency of 11.2 %. Importantly, conductive atomic force microscopy measurements on a single perovskite grain confirmed the suppression of the hysteresis due to the anti-solvent treatment, and this helps to understand the origin of the hysteresis in perovskite-based solar cells. Elucidating the effect of the anti-solvent treatment on the perovskite properties (electronic and morphology) is important and beneficial. The knowledge gained is not limited only to perovskite-based solar cells, but also to light emitting diodes and

© The Author(s) 2016
L. Etgar, *Hole Conductor Free Perovskite-based Solar Cells*,
SpringerBriefs in Applied Sciences and Technology,
DOI 10.1007/978-3-319-32991-8_8

lasing applications, which have recently involved the promising organo-metal perovskite material.

The demonstration of planar HTM free perovskite solar cell was discussed in Chap. 3. The perovskite deposition in this work was made by a special spray deposition technique, which enables changing the perovskite thickness. Perovskite thicknesses in the range of 1.4–3.4 μm were observed with micron-sized perovskite crystals depending on the number of spray passes. The best PV performance was achieved for three blocking layers with 10 spray passes having 7 % efficiency. Two main contributions to the PV mechanism were analyzed in this device structure. A depletion region was observed at the planar TiO_2/perovskite junction while charge accumulation was recognized at the perovskite/metal oxide interface.

Surprisingly, this novel, simple device structure with thick perovskite film showed reasonable PV performance even without HTM.

The ability to tune the optical properties of the OMHP was presented in Sects. 5.1 and 5.2, moreover it was demonstrated that various perovskite compositions could conduct holes efficiently and therefore could function in the HTM free configuration.

An efficient hybrid lead halide bromide/iodide perovskite hole conductor free solar cell was presented in Sect. 5.1. The hybrid perovskite was deposited by a two-step deposition technique permitting control of the perovskite composition and its band gap. Reflectance measurements using integrating sphere provide the absorption coefficient of the hybrid structures that assist in calculating the band gaps. XRD measurements show the change in the lattice parameter due to introducing the Br^- ions. The best hybrid $CH_3NH_3PbI_nBr_{3-n}$ perovskite hole conductor free solar cell achieved PCE of 8.54 % with improved stability compared to $CH_3NH_3PbI_3$ without Br substitution.

$CH(NH_2)_2PbI_3$ (Formamidinium (FA) lead halide iodide), $CH_3NH_3PbI_3$ and their mixture were also studied in HTM free solar cells, where all the perovskites studied function as light harvester and hole conductors. Surface photovoltage (SPV) measurements determined the electronic behavior and the energy gaps of the perovskites. Temperature dependence measurements show that the initial ratio between the MA and the FA cations did not necessarily define the final material composition. The change in the J_{sc} and the efficiency at various temperatures show superior stability for the $FAPbI_3$ based cells. Diffusion length of 0.8–1 μm was observed for all perovskites at a wide range of light intensities measured by Intensity modulated photovoltage/photocurrent spectroscopy (IMVS and IMPS) techniques.

One of the unique properties of the OMHP is the ability to develop high open-circuit voltage (Voc). Chapter 6 showed high voltage of 1.35 V for hole conductor free perovskite solar cells. SPV and IMVS techniques were used to elucidate the origin of the high voltage observed. The Fermi level position and the SPV spectra of the $CH_3NH_3PbI_3$ and $CH_3NH_3PbBr_3$ reveal the p-type behavior of these perovskites. The contact potential (CPD) change when light was switched on and off was higher by a factor of 2.5 for the $CH_3NH_3PbBr_3$ cells compared to the $CH_3NH_3PbI_3$ cells. The change in the CPD during illumination results with smaller surface potential for the $Al_2O_3/CH_3NH_3PbBr_3$ cells, which could contribute to the higher open-circuit voltage achieved in the $CH_3NH_3PbBr_3$ cells. Further support was observed by longer recombination lifetime for the Al_2O_3 based cells, compared to cells with mesoporous TiO_2. The high open-circuit voltage observed in cells without a hole conductor indicates that the origin of the open-circuit voltage is affected by the perovskite and the perovskite/metal oxide interface.

Finally semi-transparent perovskite-based solar cells were fabricated using simple deposition methods. The transparency of the cells was controlled by changing the opening of the mesh used for the perovskite grid assembly and by varying the concentrations of the perovskite precursor solutions. Optical microscopy, UHR-SEM, and profilometry confirmed that the perovskite layer is well confined within the grid lines. So far the best PV performance for the semi-transparent solar cells showed PCE of 4.98 % with transparency of 19 % (achieved by controlling the transparency), which is comparable to recent report on semi-transparent perovskite-based solar cell. Further improvement can be achieved by better control of additional parameters of the grid formation, such as the solvent evaporation rate, and the wetting agent type and concentration, which could result in fewer defects in the perovskite grid.